Nonlinear Physical Science

Nonlinear Physical Science

Nonlinear Physical Science focuses on recent advances of fundamental theories and principles, analytical and symbolic approaches, as well as computational techniques in nonlinear physical science and nonlinear mathematics with engineering applications.

Topics of interest in *Nonlinear Physical Science* include but are not limited to:

- New findings and discoveries in nonlinear physics and mathematics
- Nonlinearity, complexity and mathematical structures in nonlinear physics
- Nonlinear phenomena and observations in nature and engineering
- Computational methods and theories in complex systems
- Lie group analysis, new theories and principles in mathematical modeling
- Stability, bifurcation, chaos and fractals in physical science and engineering
- Nonlinear chemical and biological physics
- Discontinuity, synchronization and natural complexity in the physical sciences

Series editors

Albert C.J. Luo
Department of Mechanical and Industrial Engineering
Southern Illinois University Edwardsville
Edwardsville, IL 62026-1805, USA
e-mail: aluo@siue.edu

Nail H. Ibragimov
Department of Mathematics and Science
Blekinge Institute of Technology
S-371 79 Karlskrona, Sweden
e-mail: nib@bth.se

International Advisory Board

Ping Ao, University of Washington, USA; Email: aoping@u.washington.edu
Jan Awrejcewicz, The Technical University of Lodz, Poland; Email: awrejcew@p.lodz.pl
Eugene Benilov, University of Limerick, Ireland; Email: Eugene.Benilov@ul.ie
Eshel Ben-Jacob, Tel Aviv University, Israel; Email: eshel@tamar.tau.ac.il
Maurice Courbage, Université Paris 7, France; Email: maurice.courbage@univ-paris-diderot.fr
Marian Gidea, Northeastern Illinois University, USA; Email: mgidea@neiu.edu
James A. Glazier, Indiana University, USA; Email: glazier@indiana.edu
Shijun Liao, Shanghai Jiaotong University, China; Email: sjliao@sjtu.edu.cn
Jose Antonio Tenreiro Machado, ISEP-Institute of Engineering of Porto, Portugal; Email: jtm@dee.isep.ipp.pt
Nikolai A. Magnitskii, Russian Academy of Sciences, Russia; Email: nmag@isa.ru
Josep J. Masdemont, Universitat Politecnica de Catalunya (UPC), Spain; Email: josep@barquins.upc.edu
Dmitry E. Pelinovsky, McMaster University, Canada; Email: dmpeli@math.mcmaster.ca
Sergey Prants, V.I.Il'ichev Pacific Oceanological Institute of the Russian Academy of Sciences, Russia; Email: prants@poi.dvo.ru
Victor I. Shrira, Keele University, UK; Email: v.i.shrira@keele.ac.uk
Jian Qiao Sun, University of California, USA; Email: jqsun@ucmerced.edu
Abdul-Majid Wazwaz, Saint Xavier University, USA; Email: wazwaz@sxu.edu
Pei Yu, The University of Western Ontario, Canada; Email: pyu@uwo.ca

More information about this series at http://www.springer.com/series/8389

Marat Akhmet · Ardak Kashkynbayev

Bifurcation in Autonomous and Nonautonomous Differential Equations with Discontinuities

Marat Akhmet
Department of Mathematics
Middle East Technical University
Ankara
Turkey

Ardak Kashkynbayev
Department of Mathematics
Middle East Technical University
Ankara
Turkey

ISSN 1867-8440
Nonlinear Physical Science
ISBN 978-981-10-9809-3
DOI 10.1007/978-981-10-3180-9

ISSN 1867-8459 (electronic)

ISBN 978-981-10-3180-9 (eBook)

Jointly published with Higher Education Press, Beijing
ISBN 978-7-04-047450-3

Printed on acid-free paper

This Springer imprint is published by Springer Nature
The registered company is Springer Nature Singapore Pte Ltd.
The registered company address is: 152 Beach Road, #22-06/08 Gateway East, Singapore 189721, Singapore

To our beloved families

Preface

This book is devoted to bifurcation theory in discontinuous dynamical systems. The main novelty is the consideration of bifurcation in differential and hybrid equations by means of methods developed by authors in recent years. Hopf bifurcation results are obtained for planar and three-dimensional systems. Results in nonautonomous bifurcation theory are presented for differential equations with discontinuities. This is the first time illustrations for nonautonomous bifurcation are provided. This theory is among vast developing subjects in the recent years. The subjects in this book are evolved from:

- Bifurcation theory for autonomous and nonautononmous ODEs;
- *B*—equivalence method is developed for impulsive differential equations with nonfixed moments of impacts and principles of discontinuous dynamical systems;
- Theory of differential equations with piecewise constant argument of generalized type; and
- Theory of differential equations with discontinuous right-hand side.

We expect that the results obtained in this book will be applied to various fields such as neural networks, brain dynamics, mechanical systems, weather phenomena, and population dynamics. Thus, we think that in near future this theory will be one of the most attracting areas in dynamical systems and its applications. Without any doubt, bifurcation theory should be further developed into other types of differential equation. In this sense, we strongly believe that the present book will be a leading one in this field. Bifurcation of periodic solutions and nonautonomous systems is yet to develop in multidimensional case. Center manifold theory is one of the interesting topics to investigate.

We have published several papers and books related to bifurcation theory in recent years. In this book, we provide results in discontinuous dynamical systems that are developed parallel to ODEs. The reader will benefit from recent results obtained in the theory of bifurcation and will learn in the very concrete way how to apply this theory to differential equations with various types of discontinuity: impulsive differential equations, differential equations with piecewise constant argument, and differential equations with discontinuous right-hand side. Moreover, the reader will learn how to analyze nonautonomous bifurcation scenarios in these equations. The present book is devoted to Hopf, transcritical, and pitchfork bifurcations, and it is reasonable to discuss a new possibilities in other types of bifurcation such as Neimark–Sacker bifurcation, Shilnikov bifurcation, Bautin bifurcation, Bogdanov–Taken bifurcation, and bifurcation of almost periodic solutions.

This book will be of a big interest both for beginners and experts in the field of bifurcation theory. For the former group of specialists, that is, undergraduate and graduate students, this book will be useful since it provides strong impression that bifurcation theory can be developed not only for discrete and continuous systems but also for those which combine these systems in very different ways. The latter group of specialists will find in this book several powerful instruments developed for the theory of discontinuous dynamical systems with variable moments of impacts, differential equations with piecewise constant argument of generalized type, and Filippov systems. A significant benefit of this present book is expected to be for those who consider bifurcations in systems with discontinuities since they are presumably nonautonomous systems. Consequently, nonautonomous bifurcation is compulsory subject to discuss.

The authors would like to offer their sincere thanks to those who contributed to the preparation of this book, Duygu Aruğaslan Çinçin and Mehmet Turan for their collaboration, and the series editor Prof. Albert Luo and editor of HEP Liping Wang for their interest in the monograph and patience during the publication of the book.

Ankara, Turkey   Marat Akhmet
                                                                     Ardak Kashkynbayev

Contents

Chapter 1
Introduction

Generalizations of ordinary differential and difference equations as an abstract rule are called dynamical systems. Historically, this terminology was first used in the book of Birkhoff [71]. Although mathematical modeling of real-world processes by means of differential equations goes back to Newton it has only after Poincaré and Lyapunov, often accepted as the founders of the theory of dynamical systems, these problems started to be considered from qualitative point of view. Poincaré premised topological and geometrical approaches to analyze the dynamic behavior of the solutions instead of traditional methods known before [195–197]. On the other hand, Lyapunov in his thesis was concerned with asymptotic behavior of solutions in the neighborhood of a fixed point [169]. Thus, both of the mentioned scientists have indisputable contributions on the vast developing theory of dynamical systems as we understand and accept it today.

Poincaré used a originally French word bifurcation to explain the splitting of asymptotic behavior in a dynamical system [197]. Since then, bifurcation has been regarded as the topological change in the qualitative nature of the states as parameter varies over a specified space. These parameters often regarded as influence of an environment to a system. In autonomous dynamical systems, the bifurcation theory is well developed and it is concerned with a qualitative change in an equilibrium point or a periodic solution as parameter ranges [89, 125, 132, 148, 233]. For instance, a stable equilibrium persists as stable to small fluctuations in a certain parameter range, and as parameter crosses a critical threshold, called as a bifurcation value, the equilibrium becomes unstable or even does not exist at all. Another example is existence of a periodic solution around an equilibrium point and its disappearance in the vicinity of the bifurcation value. Hence, bifurcation is by no means exceptional but a typical property of dynamical systems such as in biochemical reactions, structural mechanics, cardiac arrhythmias in malfunctioning hearts, and in many other models of biology [89, 121, 148].

If a mathematical model explicitly involves time-dependent vector field, then it is the main object of nonautonomous dynamical systems to describe its nature. These models are usually given in terms of evolutionary equations which may be ordinary

© Springer Nature Singapore Pte Ltd. and Higher Education Press 2017
M. Akhmet and A. Kashkynbayev, *Bifurcation in Autonomous
and Nonautonomous Differential Equations with Discontinuities*,
Nonlinear Physical Science, DOI 10.1007/978-981-10-3180-9_1

differential, delay, or difference equations. We encounter with several limitations whether one assumes that an environment which surrounds system is not variable in time. The main reason for this is that conditions in real world are often very different from the ones in laboratories where models are generated. For instance, seasonal effects on different timescales or changes in nutrient supply should be taken into account when modeling predator–prey systems. Another example is to analyze possible impacts on a model after stimulating chemicals or dosing drags. Hence, there are several reasons to consider evolutionary equations with vector fields which explicitly depend on time. Statistical confirmation of this reason is often obvious since data which are obtained from a measurement may contain time-dependent parameters. There exist two approaches to study nonautonomous dynamical systems. The first one is the concept of process or two-parameter semi-flow, studied by Dafermos and Hale [99, 124]. Another approach is skew product flows which has its origins in ergodic theory which was under investigation by Sell [219, 220]. In this book, we treat process formulation to study nonautonomous dynamical systems. The classical notion of exponential dichotomy in a linear nonautonomous differential equations was introduced by Perron [188, 189] and has been under intensive research in [95, 101, 172, 210–213]. For a systematic development of nonautonomous dynamical systems in recent years, we refer to the books [139, 204].

1.1 General Description of Differential Equations with Discontinuities

The theory of differential equations with discontinuities plays an increasingly important role in applications. Many real processes which appear in various problems of biology, chemistry, control theory, ecology, economics, electronics, mechanics, medicine, and physics are studied by means of mathematical models with some kind of discontinuity [25, 26, 36, 63, 74, 100, 127, 128, 149, 164, 165, 167, 168]. This fact has increased the need to establish a comprehensive theory for differential equations with discontinuities [1, 7, 54, 56, 94, 109, 115, 122, 123, 144, 156, 173, 176, 186, 202, 215, 216].

In what follows, it is sensible to distinguish between different types of discontinuities that will be treated in this book. The first one is the discontinuous external forces also called impulse effects [1, 123, 156, 216]. Another type is the piecewise constant arguments [2, 94] of generalized type. The last one is the case when the right-hand sides of the equations depend discontinuously on the state variables [54, 115, 146]. Containing impulsive differential equations, differential equations with piecewise constant argument of generalized type and differential equations with discontinuous right-hand sides, the range of differential equations with discontinuities is quite vast.

1.1.1 Impulsive Differential Equations

There are certain cases when continuous dynamical system fail to meet the needs of real-world problems. Consider, for instance, the population dynamics of some species after epidemics or harvesting. One would expect a remarkable change in the population density of that species. Moreover, it is a known fact that there is a very quick change in momentum when a pendulum of a clock crosses its equilibrium state [59]. In addition to these, we can give as an example a rapid alter in the velocity of an oscillating string while it is struck by a hammer [144]. If one wants to be a generous, it is a necessary that in all of the above cases the mathematical models involve discontinuity. One of the most common way to study discontinuity is by means of impulsive differential equations. Although there are huge amount of literature devoted to impulsive differential equations and its applications, one in [1, 156, 216] is most commonly accepted as the fundamental work in this field. Discontinuous dynamical systems have its origin started with academic work of Krylov and Bogolyubov [144], Introduction to Nonlinear Mechanics, where the authors studied a model of a clock. This study suggested that differential equations with pulse action can be considered for nonlinear mechanics. Later, Pavlidis introduced terminology of impulsive differential equations in his studies [185, 186]. However, Samoilenko and Perestyuk in [216] developed the theory of impulsive differential equations in a more systematic way and parallel to the that of theory in ordinary differential equations. Similar ideas were used in the book of Lakshmikantham et al. [156].

The problem of investigation of differential equations with solutions, which has discontinuities on surfaces placed in (t, x)-space is one of the most difficult and interesting subjects of the theory of impulsive differential equations [4, 6, 19, 40, 52–56, 104–106, 117, 120, 131, 134, 156–158, 163, 165, 201–203, 215–217, 222, 223, 230, 234]. For this reason, it was emphasized in early stage of theories development in [176]. By using basic ideas of our predecessors [123, 215, 226], we introduce a special topology [1, 4–9, 13, 52, 53, 55, 56] in a set of piecewise continuous functions having, in general, points of discontinuity, which do not coincide. Thus, we operate with the concepts of B-topology, B-equivalence, ε-neighborhoods, when we investigate systems with variable moments of impulses [1, 4–9, 13, 52, 53, 55, 56] or consider almost periodicity of solutions of impulsive systems [9, 50, 51]. In [143], it was explored that a topology in spaces of piecewise continuous functions can be metricized. We specify this result in [51] for the investigation of almost periodic discontinuous nonautonomous systems. It is not surprising that the topology begins to be useful for other differential equations with discontinuities of different types [9, 17, 20, 28, 40, 47]: Filippov-type differential equations and differential equations on variable timescales. The most important feature of B-equivalence method is the reduction in systems with fixed moments of impulses. This method was introduced and developed in [1, 4–6, 17, 20, 39, 40, 44–52]. Let us point out that the B-equivalence method is effective not only in bounded domains, but it can also be applied successfully if impulsive equations are considered with unbounded domains [57]. Exceptionally it is important for existence of global manifolds. Linearization in the neighborhood of the nontrivial solution, the central auxiliary result of the stability theory, is solved in [4]. The problem of controllability of boundary-value problems for quasilinear impulsive system of integro-differential equations is investigated in [35]. Finally, the method also proves its effectiveness to indicate chaos and shadowing property of impulsive systems [14, 15]. Therefore, the theory of impulsive differential equations at nonfixed moments of impulses improved significantly after B-equivalence method was introduced [1, 52]. This idea enables to handle more complex systems in a coherent and fruitful way [6, 17, 20, 40, 90, 110, 162]. To be concrete, systems with impulses at variable moment of time was a hard problem to overcome over decades. There were certain attempts to solve this issue in the past; however, today it seems that B-equivalence method is an appropriate and suitable tool to address these kinds of issues. We refer to the book *Principles of Discontinuous Dynamical System* for rigorous description of impulsive systems with nonprescribed moments of discontinuity and adequate applications of B-equivalence method.

1.1.2 Differential Equations with Piecewise Constant Argument

One another way to study discontinuity in mathematical models is to consider differential equations with piecewise constant arguments. The need to study these equations raised from real-world application problems which include but not limited to the damped as well as undamped loading systems based on a piecewise constant voltage, population dynamics, neural networks, the Froude pendulum and the Geneva mechanism [2, 3, 21–23, 174, 175]. Thus, despite the fact that differential equations with piecewise constant argument is a relatively new subject, there is a vast ongoing research in this field. Nevertheless, there are very few literature which treat this theory in a systematic manner. A great step toward a characterization of the theory was achieved in [7–12, 16, 18, 22–24, 27, 28], where parallel developments similar to ordinary differential equations have been emerged. We gathered all these results in the book *Nonlinear Hybrid Continuous/Discrete Time Models* [2], which contains the basics of how this theory should be constructed. Before these results were accomplished, reduction in *discrete equations* was the main tool to treat differential equations with piecewise constant arguments. However, this method allows one to solve equations which start at integer or multiples of integer values only. This was the main obstacle to make complete qualitative analysis of a solution such as stability and bifurcation of an equilibrium. We sail through this issue by introducing an *equivalent integral equations* which permits not only to consider arbitrary piecewise constant functions as arguments but also to examine qualitative properties of a solution such as existence and uniqueness of solutions and existence of periodic and almost periodic solutions [9–12, 18, 24, 27, 29]. In other words, there is no restriction on the distance between the switching moments of the argument. Moreover, the method we propose is less restrictive since it does not require additional assumptions to reduce an equation to a discrete system. Recently, this method is widely used in the development of the theory [78–80, 87, 88, 191–193, 231]. Thus, the theory was significantly improved, and it become possible to handle more complex problems.

1.1.3 Differential Equations with Discontinuous Right-Hand Sides

It is well known that systems modeled by ordinary differential equations can be written in the vector form $x' = f(t, x)$, where $t \in \mathbb{R}$, $x \in \mathbb{R}^n$, $n \in \mathbb{N}$, and f is an n-dimensional vector-valued, continuous function. However, there exist many practical situations in which the function on the right-hand sides is discontinuous with respect to the state variable x or to the time variable t, resulting in a differential equation with discontinuous right-hand sides.

The theory of differential equations with discontinuous right-hand sides has been to a great extent developed by the needs of physical problems requiring automatic controls such as relays and switches [115]. These equations are also specific for a wide range of applications arising from mechanical systems with dry friction, electrical circuits with small inductivities, systems with small inertia, dynamical systems with nondifferentiable potential, optimization problems with nonsmooth data, electrical networks with switches, oscillations in viscoelasticity, optimal control, etc. (see, for example, [59, 115] and the references therein). Mathematical modeling of such applications leads to discontinuous systems which switch between different states, and the vector field of each state is associated with a different set of differential equations [63, 160].

Systems described by differential equations with a discontinuous right-hand sides are also called Filippov systems. For these systems, depending on the vector field, either a transversal intersection or a sliding mode may appear. From the mathematical point of view, several ways exist to handle such systems. For example, one way is to use the theory of differential inclusions [115]. Systems with sliding mode are generally extended to a set-valued vector field, that is, to differential inclusions for investigational purposes. Another way is a continuous approximation of discontinuities to get smooth differential equations [62]. Method of B-equivalence [46, 47, 54] can also be used effectively in the analysis of differential equations with discontinuous right-hand sides, especially when the sets of discontinuities are of quasilinear nature.

Stimulated by the problems of applied nature, qualitative theory of classical differential equations including the notions of existence, uniqueness, continuous dependence, stability, and bifurcation has been adapted for equations with discontinuous right-hand sides. Hence, the amount of the literature on the theory of differential equations with discontinuous right-hand sides is vast. Different aspects of the modified theory are elucidated in a variety of books and papers. The books [59, 65, 173] can be considered an important basis for the development of such systems. A nice introduction can be found in [66–68, 115, 146]. The fundamental work of Filippov extends a discontinuous differential equation to a differential inclusion [115]. In his book [115], many results from the classical theory of differential equations were shown to be valid for equations with discontinuous right-hand sides as well, and rather than applications, it presents the main trends of the theory of differential equations with discontinuous right-hand sides such as existence and uniqueness, dependence on the initial data, bounded and periodic solutions, stability, and so on. Moreover, there exist many publications that consider dry friction problems, existence and bifurcation of periodic solutions for Filippov-type systems by means of differential inclusions, see for example [63, 92, 109, 147, 150, 160, 224, 225, 236, 237]. The description of bifurcations for these systems can be found in [159, 160]. Problems in mechanical engineering may violate the requirements of smoothness if they involve collisions, finite clearances, or stick-slip phenomena. Systems of this type can display a large variety of complicated bifurcation scenarios. The book [235] contains applications of bifurcation theory to switching power converters, relay systems, and different types of pulse-width-modulated control systems.

In the literature, discontinuities on the right-hand sides are mostly assumed to appear on straight lines [92, 146, 150, 236, 237]. However, we obtained several theoretical results for such equations with nonlinear sets of discontinuities [46, 47, 54]. The main tool of investigation in these papers was the B-equivalence method introduced for differential equations with discontinuities on the right-hand sides. This method was firstly proposed to reduce impulsive systems with variable time of impulses to the systems with fixed moments of impulsive actions [55, 56]. Then, it appeared that the method is also applicable to differential equations with discontinuous right-hand sides. That is, differential equations with discontinuous vector fields with nonlinear discontinuity sets can be reduced to impulsive differential equations with fixed moments of impulses.

1.2 Nonautonomous Bifurcation

Quite simple models have played an important role in the development of the bifurcation theory in one-dimensional autonomous systems. However, it may not be appropriate to follow the same route as in autonomous dynamical systems to construct the bifurcation theory for nonautonomous dynamical systems. One of the reasons is that there may not exist an equilibrium point or a periodic solution of nonautonomous systems. Hence, most of the time the notion of equilibria is replaced with bounded trajectories. Another reason is that eigenvalues of a linearized system do not give proper information about asymptotic behavior of a solution. Thus, in scalar nonautonomous dynamical systems Lyapunov exponent seems to be an adequate tool to overcome this issue. There are strong ties between the concepts of attraction and repulsion of an invariant set and bifurcation. It is possible to predicate the base of attraction on the studies of Lyapunov [169, 170]. However, it is only in the last two decades attraction has been intensively studied in nonautonomous dynamical systems. There are basically two types of attraction in nonautonomous systems: forward and pullback. The former involves a moving invariant set and deals with attraction in Lyapunov asymptotic stability sense. The latter involves fixed invariant set which starts progressively earlier in time. The notion of pullback attractors was adapted from random dynamical systems [97, 98] and were called as cocycle attractors in some papers [140, 142]. Apparently, for the first time in the literature pullback attractor was introduced in [137] to emphasize the difference from the forward attraction. From dynamic viewpoint, there are several approaches extending bifurcation theory to nonautonomous case. The mathematical foundations of nonautonomous bifurcation theory started with the paper of Langa et al. [152], where the authors proposed to study this theory by means of pullback and forward convergence and considered the following equations as a nonautonomous counterparts of the pitchfork and saddle-node bifurcations, respectively.

$$x' = \mu_1 x - \mu_2(t)x^3,$$

and

$$x' = \mu_1 - \mu_2(t)x^2,$$

where $0 < \mu_2(t) < \mu$. Next, Langa et al. in [154] considered nonautonomous transcritical and saddle-node bifurcations and obtained sufficient conditions on Taylor coefficients of the right-hand side. Namely the authors considered the following equations.

$$x' = \lambda\mu_1(t)x - \mu_2(t)x^2$$

and

$$x' = \lambda\mu_1(t) - \mu_2(t)x^2.$$

In [108, 135], the authors obtained results for nonautonomous saddle-node and transcritical bifurcations by using the method of averaging. Nùñez and Obaya followed skew product approach in one-dimensional dynamical systems and studied bifurcation patterns depending on variation in the number and attraction properties of minimal sets [182]. One another approach is the bifurcation of control sets based on Conley index theory, which was carried out by Colonius and his coauthors [93]. Finally, there are studies which describe bifurcation in nonautonomous dynamical systems by means of attractor–repeller pair. This approach deals with the transitions of nonautonomous attractor which undergo qualitative change and become nontrivial when parameter passes through critical value [137, 141, 204]. The book of Rasmussen gives enlightening information about relation of attractivity/repulsivity and bifurcation concepts in nonautonomous systems [204].

1.3   The Bernoulli Equations

The origin of the Bernoulli differential equations go far beyond Poincaré and Lyapunov and apparently was first studied by Jacob Bernoulli in 1695 [69]. Although
these equations have already became a classical subject in the theory of differential
equations, it has not been studied, for the best of our knowledge, in the discontinuous systems yet. One of the main reasons why this subject is attractive is that by
means of the Bernoulli transformation they are reduced to a linear nonhomogeneous
equation, and hence can be solved explicitly. Despite its simplicity, recent applications in nonautonomous bifurcation theory showed that a detailed insight into the
discontinuous Bernoulli equations is necessary. We develop this simple idea to both
impulsive and hybrid systems and carry out nonautonomous bifurcation analysis as
well as analyze bounded solutions of these equations.

1.4   Organization of the Book

The remaining part of this book is organized as follows:

In Chap. 2, we start with Hopf bifurcation in two-dimensional and three-
dimensional impulsive systems. We obtain Hopf bifurcation result in discontinuous planar system and generalize this result to three-dimensional system. The jump
surfaces in this chapter are not flat. We apply the B-equivalence method to prove
the existence of discontinuous limit cycle for the van der Pol equation with impact
on surfaces. The result is extended through the center manifold theory for coupled
oscillators. Examples and simulations are provided to demonstrate the theoretical
results as well as application opportunities.

Chapter 3 is devoted to bifurcations of periodic solutions for two-dimensional
and three-dimensional nonsmooth systems. The notion of B-equivalent impulsive
systems is explained. For these systems, problems such as existence of focus and
center in the noncritical case, distinguishing between the center and the focus in the
critical case and Hopf bifurcation are solved. The center manifold theory is given
for the three-dimensional system. Appropriate examples together with numerical
simulations are presented to illustrate the findings.

In Chap. 4, we consider nonautonomous transcritical and pitchfork bifurcations
in continuous as well as discontinuous systems. The notions of so-called pullback
attractor and forward attractor are implemented to analyze asymptotic behavior of
systems. In the first part of the chapter, we study pitchfork bifurcation patterns based
on pullback convergence which depend on the properties of the system in the past.
Conditions which ensure transcritical bifurcation are obtained. In the second part of
the chapter, we not only generalize the results obtained in the first part but we also
attain less restrictive conditions to ensure nonautonomous bifurcation patters. Moreover, we introduce the Bernoulli equations in impulsive systems. The corresponding
jump equation is constructed in special way that the whole system is reduced to a

linear nonhomogeneous system under the Bernoulli transformation. Both pullback and forward asymptotic behavior of the original system is analyzed based on reduced system. In addition to these, conditions to have bounded solutions for the Bernoulli equations are achieved. Appropriate numerical simulations which illustrate theoretical results are provided.

In Chap. 5, we study nonautonomous transcritical and pitchfork bifurcations in impulsive systems which are not explicitly solvable. That is, by any means of substitution it is not possible to obtain a solution. Bifurcation scenarios in this chapter are attained in terms of qualitative change in the attractor–repeller pair. Besides, we establish new results in asymptotic behavior of linearized systems depending on entire time. In the remaining part of the chapter, finite-time analogues of nonautonomous transcritical and pitchfork bifurcations are presented in impulsive systems. Illustrative examples which support the theoretical results are depicted.

Finally, Chap. 6 is concerned with nonautonomous transcritical and pitchfork bifurcations in differential equations with alternating piecewise constant argument. The Bernoulli equation is presented for the hybrid systems. We construct special type of transformation so that the original nonlinear system is converted to a linear nonhomogeneous system. We show that bifurcation scenarios depend on the sign of Lyapunov exponents. Besides, future and past asymptotic properties of bounded solutions are discussed. Appropriate examples with numerical simulations are given to illustrate the theoretical results.

Chapter 2
Hopf Bifurcation in Impulsive Systems

2.1 Hopf Bifurcation of a Discontinuous Limit Cycle

This chapter is organized in the following manner. In the first section, we give the description of the systems under consideration and prove the theorem of existence of foci and centers of the nonperturbed system. The main subject of Sect. 2.1.2 is the foci of the perturbed equation. The noncritical case is considered. In Sect. 2.1.3, the problem of distinguishing between the center and the focus is solved. Bifurcation of a periodic solution is investigated in Sect. 2.1.4. The last section consists of examples illustrating the bifurcation theorem.

2.1.1 The Nonperturbed System

Denote by $< x, y >$ the dot product of vectors $x, y \in \mathbb{R}^2$, and $||x|| = < x, x >^{\frac{1}{2}}$, the norm of a vector $x \in \mathbb{R}^2$. Moreover, let $\mathscr{R}$ be the set of all real-valued constant 2×2 matrices, and $\mathscr{I} \in \mathscr{R}$ be the identity matrix.

D_0-**system**. Consider the following differential equation with impulses

$$\frac{dx}{dt} = Ax,$$
$$\Delta x|_{x \in \Gamma_0} = B_0 x, \tag{2.1.1}$$

where Γ_0 is a subset of $\mathbb{R}^2$, and it will be described below, $A, B_0 \in \mathscr{R}$.

The following assumptions will be needed throughout this chapter:

(C1) $\Gamma_0 = \cup_{i=1}^{p} s_i$, where p is a fixed natural number and half-lines s_i, $i = 1$, $2, \ldots, p$, are defined by equations $< a^i, x > = 0$, where $a^i = (a_1^i, a_2^i)$ are constant vectors. The origin does not belong to the lines (see Fig. 2.1).

© Springer Nature Singapore Pte Ltd. and Higher Education Press 2017
M. Akhmet and A. Kashkynbayev, *Bifurcation in Autonomous
and Nonautonomous Differential Equations with Discontinuities*,
Nonlinear Physical Science, DOI 10.1007/978-981-10-3180-9_2

Fig. 2.1 The domain of the
nonperturbed system (2.1.1)
with a vertex which unites
the straight lines s_i,
$i = 1, 2, \ldots, p$

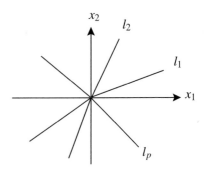

(C2)

$$A = \begin{pmatrix} \alpha & -\beta \\ \beta & \alpha \end{pmatrix},$$

where $\alpha, \beta \in \mathbb{R}$, $\beta \neq 0$;

(C3) there exists a regular matrix $Q \in \mathscr{R}$ and nonnegative real numbers k and θ
such that

$$B_0 = kQ \begin{pmatrix} \cos\theta & -\sin\theta \\ \sin\theta & \cos\theta \end{pmatrix} Q^{-1} - \begin{pmatrix} 1 & 0 \\ 0 & 1 \end{pmatrix};$$

We consider every angle for a point with respect to the positive half-line of the
first coordinate axis. Denote $s_i' = (\mathscr{I} + B_0)s_i$, $i = 1, 2, \ldots, p$. Let γ_i and ζ_i
be angles of s_i and s_i', $i = 1, 2, \ldots, p$, respectively,

$$B_0 = \begin{pmatrix} b_{11} & b_{12} \\ b_{21} & b_{22} \end{pmatrix}.$$

(C4) $0 < \gamma_1 < \zeta_1 < \gamma_2 < \cdots < \gamma_p < \zeta_p < 2\pi$, $(b_{11} + 1)\cos\gamma_i + b_{12}\sin$
$\gamma_i \neq 0$, $i = 1, 2, \ldots, p$.

If conditions (C1)–(C4) hold, then (2.1.1) is said to be a D_0-*system*.

Exercise 2.1.1 *Verify that the origin is a unique singular point of a D_0-system and
(2.1.1) is not a linear system.*

Exercise 2.1.2 *Using the results of the last chapter, prove that D_0-system (2.1.1)
provides a B-smooth discontinuous flow.*

If we use transformation $x_1 = r\cos(\phi)$, $x_2 = r\sin(\phi)$ in (2.1.1) and exclude the
time variable t, we can find that the solution $r(\phi, r_0)$ which starts at the point $(0, r_0)$,
satisfies the following system:

$$\frac{dr}{d\phi} = \lambda r,$$

$$\Delta r \mid_{\phi = \gamma_i (\mathrm{mod} 2\pi)} = k_i r, \qquad (2.1.2)$$

where $\lambda = \frac{\alpha}{\beta}$, the angle-variable ϕ is ranged over the set

$$R_\phi = \cup_{i=-\infty}^{\infty}[\cup_{j=1}^{p-1}(2\pi i + \zeta_j, 2\pi i + \gamma_{j+1}] \cup (2\pi i + \zeta_p, 2\pi(i+1) + \gamma_1]]$$

and $k_i = [((b_{11}+1)\cos(\gamma_i) + b_{12}\sin(\gamma_i))^2 + (b_{21}\cos(\gamma_i) + (b_{22}+1)\sin(\gamma_i))^2]^{\frac{1}{2}} - 1$. Equation (2.1.2) is 2π-periodic, so we shall consider just the section $\phi \in [0, 2\pi]$ in what follows. That is, the system

$$\frac{dr}{d\phi} = \lambda r,$$

$$\Delta r \mid_{\phi=\gamma_i} = k_i r, \qquad (2.1.3)$$

is considered with $\phi \in [0, 2\pi]_\phi \equiv [0, 2\pi]\setminus \cup_{i=1}^{p}(\gamma_i, \zeta_i]$. System (2.1.3) is a sample of the timescale differential equation with transition condition [38]. We shall reduce (2.1.3) to an impulsive differential equation [6, 38] for the investigation's needs. Indeed, let us introduce a new variable $\psi = \phi - \sum_{0<\gamma_j<\phi}\theta_j, \theta_j = \zeta_j - \gamma_j$, with the range $[0, 2\pi - \sum_{i=1}^{p}\theta_i]$. We shall call this new variable ψ-substitution. It is easy to check that upon ψ-substitution, the solution $r(\phi, r_0)$ satisfies the following impulsive equation

$$\frac{dr}{d\psi} = \lambda r,$$

$$\Delta r \mid_{\psi=\delta_j} = k_j r, \qquad (2.1.4)$$

where $\delta_j = \gamma_j - \sum_{0<\gamma_i<\gamma_j}\theta_i$. Solving the last impulsive system and using the inverse of ψ-substitution, one can obtain that the solution $r(\phi, r_0)$ of (2.1.2) has the form

$$r(\phi, r_0) = \exp\left(\lambda\left(\phi - \sum_{0<\gamma_i<\phi}\theta_i\right)\right)\prod_{0<\gamma_i<\phi}(1+k_i)r_0, \qquad (2.1.5)$$

if $\phi \in [0, 2\pi]_\phi$.

Denote

$$q = \exp\left(\lambda\left(2\pi - \sum_{i=1}^{p}\theta_i\right)\right)\prod_{i=1}^{p}(1+k_i). \qquad (2.1.6)$$

Applying the Poincaré return map $r(2\pi, r_0)$ to (2.1.5), one can obtain that the following theorem follows.

Theorem 2.1.1 *If*

(1) $q = 1$, then the origin is a center and all solutions of (2.1.1) are periodic with period $T = (2\pi - \sum_{i=1}^{p}\theta_i)\beta^{-1}$;

(2) $q < 1$, then the origin is a stable focus;
(3) $q > 1$, then the origin is an unstable focus of D_0-system.

2.1.2 The Perturbed System

Theorem 2.1.1 of the last section implies that if conditions $(C1)$–$(C4)$ are valid, then each trajectory of (2.1.1) either spirals to the origin or is a discontinuous cycle. Moreover, if the trajectory spirals to the origin, then it spirals to infinity, too. That is, the asymptotic behavior of the trajectory is very similar to the behavior of trajectories of the planar linear system of ordinary differential equations with constant coefficients [91, 126]. In what follows, we will consider how a perturbation may change the phase portrait of the system.

D-**system.** Let us consider the following equation:

$$\frac{dx}{dt} = Ax + f(x),$$
$$\Delta x|_{x \in \Gamma} = B(x)x, \tag{2.1.7}$$

in a neighborhood G of the origin.

The following is the list of conditions assumed for this system:

(C5) $\Gamma = \cup_{i=1}^{p} l_i$ is a set of curves which start at the origin and are determined by the equations $< a^i, x > +\tau_i(x) = 0, \quad i = 1, 2, \ldots, p$. The origin does not belong to the curves (see Fig. 2.2).

(C6)

$$B(x) = (k + \kappa(x)) Q \begin{pmatrix} \cos(\theta + \upsilon(x)) & -\sin(\theta + \upsilon(x)) \\ \sin(\theta + \upsilon(x)) & \cos(\theta + \upsilon(x)) \end{pmatrix} Q^{-1} - \begin{pmatrix} 1 & 0 \\ 0 & 1 \end{pmatrix},$$

$(\mathscr{I} + B(x))x \in G$ for all $x \in G$;

(C7) $\{f, \kappa, \upsilon\} \subset C^{(1)}(G), \{\tau_i, i = 1, 2, \ldots, p\} \subset C^{(2)}(G)$;

Fig. 2.2 The domain of the perturbed system (2.1.7) near a vertex which unites the curves l_i associated with the straight lines s_i, $i = 1, 2, \ldots, p$

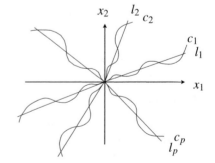

(C8) $f(x) = o(||x||)$, $\kappa(x) = o(||x||)$, $\upsilon(x) = o(||x||)$, $\tau_i(x) = o(||x||^2)$, $i = 1$, $2, \ldots, p$;

Moreover, we assume that the matrices A, Q, the vectors a^i, $i = 1, 2, \ldots, p$, and constants k, θ are the same as in (2.1.1); i.e.,

(C9) The system (2.1.1) is D_0-system associated with (2.1.7).

If conditions $(C1)$–$(C9)$ hold, then the system (2.1.7) is said to be a D-system. If G is sufficiently small, then conditions $(C4)$ and $(C8)$ imply that none of curves l_i intersect itself, they do not intersect each other, and the origin is a unique singular point of the D-system.

Exercise 2.1.3 *Using the results of the last chapter, and Example 2.1.2, prove that D-system defines a B-smooth discontinuous flow.*

Assume, without loss of generality, that $\gamma_i \neq \frac{\pi}{2} j$, $j = 1, 3$, and transform the equations in $(C5)$ to the polar coordinates so that $l_i : a_i^1 r \cos(\phi) + a_i^2 r \sin(\phi) + \tau_i(r \cos(\phi), r \sin(\phi)) = 0$ or

$$\phi = \tan^{-1}\left(\tan \gamma_i - \frac{\tau_i}{a_i^2 r \cos(\phi)}\right).$$

Now, use Taylor's expansion to get that

$$l_i : \phi = \gamma_i + r \psi_i(r, \phi), \tag{2.1.8}$$

$i = 1, 2, \ldots, p$, where ψ_i are 2π-periodic in ϕ, continuously differentiable functions, and $\psi_i = O(r)$. If the point $x(t)$ meets the discontinuity curve l_i with an angle θ, then the point $x(\theta+)$ belongs to the curve $l_i' = \{z \in \mathbb{R}^2 | z = (\mathscr{I} + B(x))x, x \in l_i\}$. The following assertion is very important for the rest of the chapter.

Lemma 2.1.1 *Suppose $(C7)$ and $(C8)$ are satisfied. Then the curve l_i', $1 \leq i \leq p$, is placed between l_i and l_{i+1}, if G is sufficiently small.*

Proof Fix $i = 1, 2, \ldots, p$, and assume that s_i, s_{i+1}, l_i, l_{i+1} are transformed by the map $y = Q^{-1}x$ into lines s_i'', s_{i+1}'', l_i'', l_{i+1}'', respectively. Set $L_i = \{z \in \mathbb{R}^2 | z = Q^{-1}(I + B(Qy))Qy, y \in l_i''\}$, $\xi_i = Q^{-1}(I + B_0)Qs_i''$, and let γ_i'', γ_{i+1}'', ζ_i' be the angles of straight lines s_i'', s_{i+1}'', ξ_i. We may assume, without loss of generality, that $\gamma_i' < \zeta_i' < \gamma_{i+1}'$. To prove the lemma, it is sufficient to check whether L_i lies between curves l_i'', l_{i+1}''. Suppose that $0 < \gamma_i' < \zeta_i' < \gamma_{i+1}' < \frac{\pi}{2}$. Otherwise, one can use a linear transformation, which does not change the relation of the curves. Let $c_1 y_1 + c_2 y_2 + l^*(y_1, y_2) = 0$ be the equation of the line l_i''. Use the polar coordinates $y_1 = \rho \cos(\phi)$, $y_2 = \rho \sin(\phi)$ and obtain $\phi = \gamma_i' + \rho \psi^*(\rho, \phi)$, where $\psi^*(\rho, \phi) = O(\rho)$ and ψ^* is a 2π-periodic function. If $y = (y_1, y_2) \in l_i''$, then the point

$$y^+ = Q^{-1}(B(Qy) + I)Qy, \tag{2.1.9}$$

where $y^+ = (y_1^+, y_2^+)$ belongs to L_i. Assume without loss of generality that $y_1^+ \neq 0$. Otherwise, use the condition $y_2^+ \neq 0$. If we set $\rho = (y_1^2 + y_2^2)^{\frac{1}{2}}, \phi = \tan^{-1}(\frac{y_2}{y_1}), \rho^+ = ((y_1^+)^2 + (y_2^+)^2)^{\frac{1}{2}}, \phi^+ = \tan^{-1}(\frac{y_2^+}{y_1^+})$, then (2.1.9) implies that

$$\rho^+ = k_i \rho + \rho \beta^*(\rho, \phi), \tag{2.1.10}$$

$$\phi^+ = \phi + \theta + \gamma^*(\rho, \phi), \tag{2.1.11}$$

where β^* and γ^* are 2π-periodic in ϕ functions and $\beta^* = O(\rho), \gamma^* = O(\rho)$. Let $\sigma(y_1, y_2) = c_1 y_1 + c_2 y_2 + l^*(y_1, y_2)$. Then,

$$\sigma(y_1^+, y_2^+) = \rho^+(c_1 \cos(\phi^+) + c_2 \sin(\phi^+) + l^*(\rho^+ \cos(\phi^+), \rho^+ \sin(\phi^+)) =$$

$$\rho^+ \sqrt{c_1^2 + c_2^2} \sin(\theta + \upsilon(\rho, \phi) - \rho\psi^*(\rho, \psi)) + l^*(\rho^+ \cos(\phi^+), \rho^+ \sin(\phi^+)),$$

where $\upsilon(\rho, \phi) = \upsilon(Qy)$. It is readily seen that the sign of $\sigma(\rho^+, \phi^+)$ is the same as of $\sin(\theta)$, if ρ is sufficiently small. Consequently, $\sigma(\rho^+, \phi^+) > 0$. Thus, the curve L_i is placed above the curve l_i'' in the first quarter of the plane Ox_1x_2. Similarly, one can show that it is placed below l_{i+1}''. The lemma is proved.

The last lemma guarantees that if G is sufficiently small, then every nontrivial trajectory of the system (2.1.7) meets each of the lines $l_i, i = 1, 2, \ldots, p$, precisely once within any time interval of length T.

2.1.3 Foci of the D-System

Utilize the polar coordinates $x_1 = r \cos(\phi), x_2 = r \sin(\phi)$ to reduce the differential part of (2.1.7) to the following form:

$$\frac{dr}{d\phi} = \lambda r + P(r, \phi).$$

It is known [60, 91, 173, 180] that $P(r, \phi)$ is 2π-periodic, continuously differentiable function, and $P = o(r)$. Set $x^+ = (x_1^+, x_2^+) = (\mathscr{I} + B(x))x, \quad x^+ = r^+(\cos \phi^+, \sin \phi^+), \tilde{x}^+ = (\tilde{x}_1^+, \tilde{x}_2^+) = (\mathscr{I} + B(0))x$, where $x = (x_1, x_2) \in l_i, \ i = 1, 2, \ldots, p$. One can find that the inequality $||x^+ - \tilde{x}^+|| \leq ||B(x) - B(0)||||x||$ implies $r^+ = r + k_i r + \omega(r, \phi)$. Use the relation between $\frac{x_2^+}{x_1^+}$ and $\frac{\tilde{x}_2^+}{\tilde{x}_1^+}$ and condition (C5) to obtain that $\phi^+ = \phi + \theta_i + \gamma(r, \phi)$. Functions ω, γ are 2π-periodic in ϕ and $\omega = o(r), \gamma(r, \phi) = o(r)$. Finally, (2.1.7) has the form

$$\frac{dr}{d\phi} = \lambda r + P(r, \phi),$$

$$\Delta r \mid_{(\rho, \phi) \in l_i} = k_i r + \omega(r, \phi),$$

$$\Delta \phi \mid_{(\rho, \phi) \in l_i} = \theta_i + \gamma(r, \phi). \tag{2.1.12}$$

It is convenient to introduce the following version of B-equivalence.
Introduce the following system:

$$\frac{d\rho}{d\phi} = \lambda \rho + P(\rho, \phi),$$

$$\Delta \rho \mid_{\phi = \gamma_i} = k_i \rho + w_i(\rho),$$

$$\Delta \phi \mid_{\phi = \gamma_i} = \theta_i, \tag{2.1.13}$$

where all elements, except w_i, $i = 1, 2, \ldots, p$, are the same as in (2.1.12) and the domain of (2.1.13) is $[0, 2\pi]_\phi$. Functions w_i will be defined below.

Let $r(\phi, r_0)$, $r(0, r_0) = r_0$, be a solution of (2.1.12) and ϕ_i be the angle where the solution intersects l_i. Denote by $\chi_i = \phi_i + \theta_i + \gamma(r(\phi_i, r_0), \phi_i)$ the angle of $r(\phi, r_0)$ after the jump.

We shall say that systems (2.1.12) and (2.1.13) are B-equivalent in G if there exists a neighborhood $G_1 \subset G$ of the origin such that for every solution $r(\phi, r_0)$ of (2.1.12) whose trajectory is in G_1, there exists a solution $\rho(\phi, r_0)$, $\rho(0, r_0) = r_0$, of (2.1.13) which satisfies the relation

$$r(\phi, r_0) = \rho(\phi, r_0), \phi \in [0, 2\pi]_\phi \setminus \cup_{i=1}^{p} \{[\phi_i, \hat{\gamma_i},] \cup [\zeta_i, \hat{\chi_i}]\}, \tag{2.1.14}$$

and, conversely, for every solution $\rho(\phi, r_0)$ of (2.1.13) whose trajectory is in G_1, there exists a solution $r(\phi, r_0)$ of (2.1.12) which satisfies (2.1.14).

We will define functions w_i such that systems (2.1.12) and (2.1.13) are B-equivalent in G, if the domain is sufficiently small.

Fix i. Let $r_1(\phi, \gamma_i, \rho)$, $r_1(\gamma_i, \gamma_i, \rho) = \rho$, be a solution of the equation

$$\frac{dr}{d\phi} = \lambda r + P(r, \phi) \tag{2.1.15}$$

and $\phi = \eta_i$ be the meeting angle of $r_1(\phi, \gamma_i, \rho)$ with l_i. Then,

$$r_1(\eta_i, \gamma_i, \rho) = \exp(\lambda(\eta_i - \gamma_i))\rho + \int_{\gamma_i}^{\eta_i} \exp(\lambda(\eta_i - s))P(r_1(s, \gamma_i, \rho), s)ds.$$

Let $\eta_i^1 = \eta_i + \theta_i + \gamma(r_1(\eta_i, \gamma_i, \rho), \eta_i)$, $\rho^1 = (1 + k_i)r_1(\eta_i, \gamma_i, \rho) + \omega(r(\eta_i, \gamma_i, \rho), \eta_i)$, and $r_2(\phi, \eta_i^1, \rho^1)$ be the solution of system (2.1.15),

$$r_2(\zeta_i, \eta_i^1, \rho^1) = \exp(\lambda(\zeta_i - \eta_i^1))\rho^1 + \int_{\eta_i^1}^{\zeta_i} \exp(\lambda(\zeta_i - s))P(r_2(s, \eta_i^1, \rho^1), s)ds.$$

Introduce

$$w_i(\rho) = r_2(\zeta_i, \eta_i^1, \rho^1) - (1 + k_i)\rho = \exp(\lambda(\zeta_i - \eta_i^1))[(1 + k_i)(\exp(\lambda(\eta_i - \gamma_i))\rho +$$

$$\int_{\gamma_i}^{\eta_i} \exp(\lambda(\eta_i - s))P(r_1(s, \gamma_i, \rho), s)ds) + \omega(r_1(\eta_i, \gamma_i, \rho), \eta_i)] +$$

$$\int_{\eta_i^1}^{\zeta_i} \exp(\lambda(\zeta_i - s))P(r_2(s, \eta_i^1, \rho^1), s)ds - (1 + k)\rho$$

or, if simplified,

$$w_i(\rho) = (1 + k)[\exp(-\lambda\gamma(r_1(\eta_i, \gamma_i, \rho), \eta_i)) - 1]\rho +$$

$$(1 + k)\int_{\gamma_i}^{\eta_i} \exp(\lambda(\zeta_i - \theta_i - s - \rho\gamma(r_1(\eta_i, \gamma_i, \rho), \eta_i)))P(r_1(s, \gamma_i, \rho), s)ds +$$

$$\int_{\eta_i^1}^{\zeta_i} \exp(\lambda(\zeta_i - s))P(r_2(s, \eta_i^1, \rho^1), s)ds +$$

$$\exp(\lambda(\zeta_i - \eta_i^1))\omega(r_1(\eta_i, \gamma_i, \rho), \eta_i). \tag{2.1.16}$$

Differentiating (2.1.8) and (2.1.16), one can find that

$$\frac{d\eta_i}{d\rho} = \frac{\frac{\partial r_1}{\partial \rho}[\psi_i + r_1\frac{\partial \psi_i}{\partial r}]}{1 - (\lambda r_1 + P)[\psi_i + r_1\frac{\partial \psi_i}{\partial r}] - r_1\frac{\partial \psi_i}{\partial \phi}}, \quad \frac{d\eta_i^1}{d\rho} = \frac{d\eta_i}{d\rho}\left(1 + \frac{\partial\gamma}{\partial\phi}\right) + \frac{\partial\gamma}{\partial r}\frac{\partial r_1}{\partial\rho},$$

$$\frac{dw_i}{d\rho} = (1 + k_i)[e^{-\lambda\gamma} - 1] - \lambda(1 + k_i)e^{-\lambda\gamma}\left(\frac{\partial\gamma}{\partial r}\frac{\partial r_1}{\partial\rho} + \frac{\partial\gamma}{\partial\phi}\frac{d\eta_i}{d\rho}\right)\rho +$$

$$(1 + k_i)e^{\lambda(\zeta_i - \theta_i - \eta_i - \gamma)}P\frac{d\eta_i}{d\rho} +$$

$$(1 + k_i)\int_{\gamma_i}^{\eta_i} e^{\lambda(\zeta_i - \theta - s - \gamma)}\left\{-\lambda\left(\frac{\partial\gamma}{\partial r}\frac{\partial r_1}{\partial\rho} + \frac{\partial\gamma}{\partial\phi}\frac{d\eta_i}{d\rho}\right)P - \frac{\partial P}{\partial r}\frac{\partial r_1}{\partial\rho} - \frac{\partial P}{\partial\phi}\frac{d\eta_i}{d\rho}\right\}ds +$$

$$\int_{\eta_i^1}^{\zeta_i} e^{\lambda(\zeta_i - s)}\frac{\partial P(r_2(s, \eta_i^1, \rho^1), s)}{\partial r}\frac{\partial r_2}{\partial\rho}ds - e^{\lambda(\zeta_i - \eta_i^1)}P(\rho^1, \eta_i^1)\frac{\partial\eta_i^1}{\partial\rho} +$$

$$e^{\lambda(\zeta_i - \eta_i^1)}\left[-\frac{\partial\eta_i^1}{\partial\rho}\omega + \frac{\partial\omega}{\partial r}\frac{\partial r_1}{\partial\rho} + \frac{\partial\omega}{\partial\phi}\frac{d\eta_i}{d\rho}\right]. \tag{2.1.17}$$

Analyzing (2.1.16) and (2.1.17), one can prove that the following two lemmas are valid.

Lemma 2.1.2 *If conditions (C1)–(C5) are valid then w_i is a continuously differentiable function, and $w_i(\rho) = o(\rho), i = 1, 2, \ldots, p$.*

Lemma 2.1.3 *The systems (2.1.12) and (2.1.13) are B-equivalent if G is sufficiently small.*

Theorem 2.1.2 *Suppose that* $(C1)$–$(C6)$ *are satisfied and* $q < 1$ $(q > 1)$. *Then the origin is a stable (unstable) focus of system* $(2.1.7)$.

Proof Let $r(\phi, r_0)$, $r(0, r_0) = r_0$, be the solution of $(2.1.12)$, and $\rho(\phi, r_0)$, $\rho(0, r_0) = r_0$, be the solution of $(2.1.13)$. Using ψ-substitution, one can obtain that

$$
\begin{aligned}
\rho(\phi, r_0) = \exp(\lambda\phi) \Bigg\{ &\Pi_{i=1}^{m}(1 + k_i)\exp\left(-\lambda\sum_{s=1}^{m}\theta_s\right)r_0 + \\
&\Pi_{i=1}^{m}(1 + k_i)\exp\left(-\lambda\sum_{s=1}^{m}\theta_s\right)\int_{0}^{\gamma_1}\exp(-\lambda u)P\,du + \\
&\Pi_{i=2}^{m}(1 + k_i)\exp\left(-\lambda\sum_{s=2}^{m}\theta_s\right)\int_{\zeta_1}^{\gamma_2}\exp(-\lambda u)P\,du + \cdots \\
&\int_{\zeta_m}^{\phi}\exp(-\lambda u)P\,du + \Pi_{i=2}^{m}(1 + k_i)\exp\left(-\lambda\sum_{s=2}^{m}\theta_s\right)w_1 + \\
&\Pi_{i=3}^{m}(1 + k_i)\exp\left(-\lambda\sum_{s=3}^{m}\theta_s\right)w_2 \ldots + \exp(-\lambda\zeta_m)w_m \Bigg\},
\end{aligned}
\tag{2.1.18}
$$

where $\phi \in [0, 2\pi]_\phi$, $P = P(\rho(\phi, r_0), \phi)$, $w_i = w_i(\rho(\gamma_i, r_0))$. Now, applying Theorem 6.1.1 in [1], conditions $(C4)$, $(C5)$ and Lemma 2.1.2, one can find that the solution $\rho(\psi, r_0)$ is differentiable in r_0 and the derivative $\frac{\partial\rho(\phi, r_0)}{\partial r_0}$ at the point $(2\pi, 0)$ is equal to q. Since $(2.1.12)$ and $(2.1.13)$ are B-equivalent, it follows that

$$
\frac{\partial r(2\pi, 0)}{\partial r_0} = q
$$

and the proof is completed.

2.1.4 The Center and Focus Problem

Throughout this section, we assume that $q = 1$. That is, the critical case is considered. Functions $f, \kappa, v, \tau_i, i = 1, 2, \ldots, p$, are assumed to be analytic in G. By condition $(C8)$, Taylor's expansions of functions f, κ, and v start with members of order not less than 2, and the expansions of $\tau_i, i = 1, 2, \ldots, p$, start with members of order not less than 3. First, we investigate the problem for $(2.1.13)$ all of whose elements are analytic functions, if ρ is sufficiently small. Theorem 6.4.2 in [1] implies that $w_i, i = 1, 2, \ldots, p$, are analytic functions in ρ and the solution $\rho(\phi, r_0)$ of equation $(2.1.13)$ has the following expansion:

$$\rho(\phi, r_0) = \sum_{i=0}^{\infty} \rho_i(\phi) r_0^i, \tag{2.1.19}$$

where $\phi \notin (\gamma_i, \zeta_i], i = 1, 2, \ldots, p, \rho_0(\phi) = 0, q = \rho_1(\phi) = 1$. One can define the Poincaré return map

$$\rho(2\pi, r_0) = \sum_{i=1}^{\infty} a_i r_0^i, \tag{2.1.20}$$

where $a_i = \rho_i(2\pi), i \geq 1, a_1 = q = 1$. The expansions exist, see Sect. 6.4 of the book [1], such that

$$P(\rho, \phi) = \sum_{i=2}^{\infty} P_i(\phi) \rho^i,$$

$$w_j(\rho) = \sum_{i=2}^{\infty} w_{ji} \rho^i, \tag{2.1.21}$$

where $P_i(\phi), w_{ji}(\phi), j \geq 2$, are 2π-periodic functions which can be defined by using (2.1.12). The coefficient $\rho_j(\phi), j \geq 2$, is the solution of the system

$$\frac{d\rho}{d\phi} = P_j(\phi),$$

$$\Delta\rho \mid_{\phi \neq \gamma_i} = w_{ji},$$

$$\Delta\phi \mid_{\phi \neq \gamma_i} = \theta_i, \tag{2.1.22}$$

with the initial condition $\rho_j(0) = 0$. Hence, coefficients of (2.1.20) are equal to

$$a_j = \int_0^{\gamma_1} P_j(\phi) d\phi + \sum_{i=1}^{p-1} \int_{\zeta_i}^{\gamma_{i+1}} P_j(\phi) d\phi + \int_{\zeta_p}^{2\pi} P_j(\phi) d\phi + \sum_{i=1}^{p} w_{ji}. \tag{2.1.23}$$

From (2.1.20) and (2.1.23), it follows that the following lemma is true.

Lemma 2.1.4 *Let $q = 1$ and the first nonzero element of the sequence $a_j, j \geq 2$, be negative (positive), then the origin is a stable (unstable) focus of (2.1.13). If $a_j = 0, j \geq 2$, then the origin is a center of (2.1.13).*

B-equivalence of systems (2.1.12) and (2.1.13) implies immediately that the following theorem is valid.

Theorem 2.1.3 *Let $q = 1$ and the first nonzero element of the sequence $a_j, j \geq 2$, be negative (positive), then the origin is a stable (unstable) focus of the equation (2.1.7). If $a_j = 0$ for all $j \geq 2$, then the origin is a center of (2.1.7).*

2.1.5 Bifurcation of a Discontinuous Limit Cycle

We consider the following system:

$$\frac{dx}{dt} = Ax + f(x) + \mu F(x, \mu),$$
$$\Delta x|_{x \in \Gamma(\mu)} = B(x, \mu)x. \tag{2.1.24}$$

To establish the Hopf bifurcation theorem, we need the following assumptions:

(A1) The set $\Gamma(\mu) = \cup_{i=1}^{p} l_i(\mu)$ is a union of curves in G, which start at the origin and do not include it, $l_i : (a^i, x) + \tau_i(x) + \mu v(x, \mu) = 0, 1 \leq i \leq p$;

(A2) There exist a matrix $Q(\mu) \in \mathscr{R}, Q(0) = Q$, analytic in $(-\mu_0, \mu_0)$, and real numbers γ, χ such that $Q^{-1}(\mu)B(x, \mu)Q(\mu) =$

$$(k + \mu\gamma + \kappa(x)) \begin{pmatrix} \cos(\theta + \mu\chi + \upsilon(x)) & -\sin(\theta + \mu\chi + \upsilon(x)) \\ \sin(\theta + \mu\chi + \upsilon(x)) & \cos(\theta + \mu\chi + \upsilon(x)) \end{pmatrix} - \begin{pmatrix} 1 & 0 \\ 0 & 1 \end{pmatrix};$$

(A3) Associated with (2.1.24) systems

$$\frac{dx}{dt} = Ax,$$
$$\Delta x|_{x \in \Gamma(0)} = B_0 x, \tag{2.1.25}$$

and

$$\frac{dx}{dt} = Ax + f(x),$$
$$\Delta x|_{x \in \Gamma(0)} = B(x, 0)x, \tag{2.1.26}$$

are D_0-system and D-system, respectively;

(A4) Functions $\kappa, \upsilon : G \to \mathbb{R}^2$ and $F, v : G \times (-\mu_0, \mu_0) \to \mathbb{R}^2$ are analytic in $G \times (-\mu_0, \mu_0)$;

(A5) $F(0, \mu) = 0, v(0, \mu) = 0$, for all $\mu \in (-\mu_0, \mu_0)$.

Additionally, we shall need the following system:

$$\frac{dx}{dt} = A(\mu)x,$$
$$\Delta x|_{x \in \Gamma_0(\mu)} = B(0, \mu)x, \tag{2.1.27}$$

where $A(\mu) = A + \mu \frac{\partial F(0, \mu)}{\partial x}$, and $\Gamma_0(\mu) = \cup_{i=1}^{p} m_i$ with

$$m_i : \quad \left(a^i + \mu \frac{\partial v(0, \mu)}{\partial x}, x \right) = 0, \quad i = 1, 2, \ldots, p.$$

The polar transformation takes (2.1.24) to the following form:

$$\frac{dr}{d\phi} = \lambda r + P(r, \phi, \mu),$$

$$\Delta r \mid_{(r,\phi)\in l_i(\mu)} = k_i r + \omega(r, \phi, \mu),$$

$$\Delta\phi \mid_{(r,\phi)\in l_i(\mu)} = \theta_i + r\gamma(r, \phi, \mu). \qquad (2.1.28)$$

The functions $w_i(\rho, \mu)$ can be defined in the same manner as in (2.1.16) such that the system

$$\frac{d\rho}{d\phi} = \lambda\rho + P(\rho, \phi, \mu), \ \phi \neq \gamma_i(\mu),$$

$$\Delta\rho \mid_{\phi=\gamma_i(\mu)} = k_i \rho + w_i(\rho, \mu),$$

$$\Delta\phi \mid_{\phi=\gamma_i(\mu)} = \theta_i(\mu), \qquad (2.1.29)$$

where $\gamma_i(\mu)$, $i = 1, 2, \ldots, p$, are angles of m_i, is B-equivalent to (2.1.28).

Similar to (2.1.6), one can define the function

$$q(\mu) = \exp(\lambda(\mu)(2\pi - \sum_{j=1}^{p}(\zeta_j(\mu) - \gamma_j(\mu))\Pi_{j=p}^{1}(1 + k_j(\mu)) \quad (2.1.30)$$

for system (2.1.27). Theorem 6.4.2 of Chap. 6 in [1] implies that $q(\mu)$ is an analytic function.

Theorem 2.1.4 *Assume that $q(0) = 1$, $q'(0) \neq 0$ and the origin is a focus of (2.1.26). Then, for sufficiently small r_0, there exists a continuous function $\mu = \delta(r_0)$, $\delta(0) = 0$, such that the solution $r(\phi, r_0, \delta(r_0))$ of (2.1.28) is periodic function with period 2π. The period of the corresponding solution of (2.1.24) is $T = (2\pi - \sum_{i=1}^{p} \theta_i)\beta^{-1} + o(|\mu|)$. Moreover, if the origin is a stable focus of (2.1.26) then the closed trajectory is a limit cycle.*

Proof If $\rho(\phi, r_0, \mu)$ is a solution of (2.1.29), then by Theorem 6.4.2 in [1] we have that

$$\rho(2\pi, r_0, \mu) = \sum_{i=1}^{\infty} a_i(\mu)r_0^i,$$

where $a_i(\mu) = \sum_{j=0}^{\infty} a_{ij}\mu^j$, $a_{10} = q(0) = 1$, $a_{11} = q'(0) \neq 0$. Define the displacement function

$$\mathscr{V}(r_0, \mu) = \rho(2\pi, r_0, \mu) - r_0 = q'(0)\mu r_0 + \sum_{i=2}^{\infty} a_{i0}r_0^i + r_0\mu^2 G_1(r_0, \mu) + r_0^2\mu G_2(r_0, \mu),$$

where G_1, G_2 are functions analytic in a neighborhood of $(0, 0)$. The bifurcation equation is $\mathscr{V}(r_0, \mu) = 0$. Canceling by r_0, one can rewrite the equation as

$$\mathscr{H}(r_0, \mu) = 0, \tag{2.1.31}$$

where

$$\mathscr{H}(r_0, \mu) = q'(0)\mu + \sum_{i=2}^{\infty} a_{i0} r_0^{i-1} + \mu^2 G_1(r_0, \mu) + r_0 \mu G_2(r_0, \mu)$$

Since

$$\mathscr{H}(0, 0) = 0, \quad \frac{\partial \mathscr{H}(0, 0)}{\partial \mu} = q'(0) \neq 0,$$

for sufficiently small r_0, there exists a function $\mu = \delta(r_0)$ such that $r(\phi, r_0, \delta(r_0))$ is a periodic solution. If conditions $a_{i0} = 0, i = 2, \ldots, l - 1$, and $a_{l0} \neq 0$ are valid, then one can obtain from (2.1.31) that

$$\delta(r_0) = -\frac{a_{l0}}{q'(0)} r_0^{l-1} + \sum_{i=l}^{\infty} \delta_i r_0^i. \tag{2.1.32}$$

By analysis of the latter expression, one can conclude that the bifurcation of periodic solutions emerges if the focus is stable with $\mu = 0$ and unstable with $\mu \neq 0$ and conversely. If $\rho(\phi) = \rho(\phi, \bar{r}_0, \bar{\mu})$ is a periodic solution of (2.1.29), then it is known that the trajectory is a limit cycle if

$$\frac{\partial \mathscr{V}(\bar{r}_0, \bar{\mu})}{\partial r_0} < 0. \tag{2.1.33}$$

We have that

$$\frac{\partial \mathscr{V}(r_0, \mu)}{\partial r_0} = q'(0)\mu + \sum_{i=2}^{\infty} i a_{i0} r_0^{i-1} + \mu^2 G_1(r_0, \mu) + 2r_0 \mu G_2(r_0, \mu).$$

Let a_{l0} be the first nonzero element among a_{i0} and $a_{l0} < 0$. Using (2.1.32), one can obtain that

$$\frac{\partial \mathscr{V}(\bar{r}_0, \bar{\mu})}{\partial r_0} = (l - 1) a_{l0} \bar{r}_0^{l-1} + Q(\bar{r}_0),$$

where Q starts with a member whose order is not less than l. Hence, (2.1.33) is valid. Now, B-equivalence of (2.1.28) and (2.1.29) proves the theorem.

Fig. 2.3 A Hopf bifurcation
diagram of an ordinary
differential equation

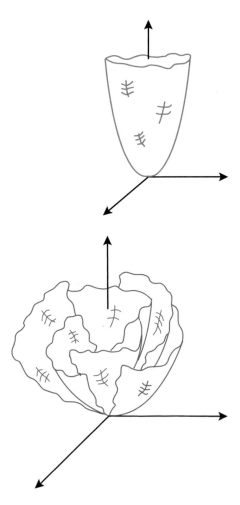

Fig. 2.4 A Hopf bifurcation
diagram of a discontinuous
dynamical system

Remark 2.1.1 (*a*). It is important to notice that the bifurcation theorem can be
obtained by applying the results in [132] and theorems of Chap. 6 of [1]. We follow the approach which is focused on the expansions of solutions [173].

(*b*). To illustrate that discontinuous dynamical systems may provide more interesting opportunities than continuous dynamics, let us compare the bifurcation diagrams
of an ordinary differential equation, Fig. 2.3, and a discontinuous dynamical system
of type (2.1.24), Fig. 2.4. One can see that the first diagram resembles a bud, and the
second one a rose. They demonstrate that a theory of differential equations flourishes
if a discontinuity is involved in analysis.

2.1.6 Examples

Example 2.1.1 Consider the following system

$$x_1' = (2 + \mu)x_1 - x_2 + x_1^2 x_2,$$
$$x_2' = x_1 + (2 + \mu)x_2 + 3x_1^3 x_2,$$
$$\Delta x_1|_{x \in l} = \left((\kappa + \mu^2)\cos\left(\frac{\pi}{6}\right) - 1\right)x_1 - (\kappa + \mu^2)\sin\left(\frac{\pi}{6}\right)x_2,$$
$$\Delta x_2|_{x \in l} = (\kappa + \mu^2)\sin\left(\frac{\pi}{6}\right)x_1 + \left((\kappa + \mu^2)\cos\left(\frac{\pi}{6}\right) - 1\right)x_2, \quad (2.1.34)$$

where $\kappa = e^{-\frac{11\pi}{6}}$, and the curve l is given by the equation $x_2 = x_1^3$, where $x_1 > 0$. One can define, using (2.1.30), that $q(\mu) = (\kappa + \mu^2)\exp((2 + \mu)\frac{11\pi}{6})$, $q(0) = \kappa \exp(\frac{11\pi}{3}) = 1$, $q'(0) = -\frac{11\pi}{6} \neq 0$. Thus, by Theorem 2.1.4, system (2.1.34) has a periodic solution with period $\approx \frac{11\pi}{12}$ if $|\mu|$ is sufficiently small.

Example 2.1.2 Let the following system be given

$$x_1' = (\mu - 1)x_1 - x_2, \quad x_2' = x_1 + (\mu - 1)x_2,$$
$$\Delta x_1|_{x \in l} = \left((\kappa - x_1^2 - x_2^2)\cos\left(\frac{\pi}{4}\right) - 1\right)x_1 - (\kappa - x_1^2 - x_2^2)\sin\left(\frac{\pi}{4}\right)x_2,$$
$$\Delta x_2|_{x \in l} = (\kappa - x_1^2 - x_2^2)\sin\left(\frac{\pi}{4}\right)x_1 + \left((\kappa - x_1^2 - x_2^2)\cos\left(\frac{\pi}{4}\right) - 1\right)x_2,$$
$$(2.1.35)$$

where l is a curve given by the equation $x_2 = x_1 + \mu x_1^2$, $x_1 > 0$, $\kappa = \exp(\frac{7\pi}{4})$. Using (2.1.30) one can find that $q(\mu) = \kappa \exp((\mu - 1)\frac{7\pi}{4})$, $q(0) = \kappa \exp(-\frac{7\pi}{4}) = 1$, $q'(0) = \frac{7\pi}{4} \neq 0$. Moreover, one can see that for the associated D-system

$$x_1' = -x_1 - x_2, \quad x_2' = x_1 - x_2,$$
$$\Delta x_1|_{x \in s} = \left((\kappa - x_1^2 - x_2^2)\cos\left(\frac{\pi}{4}\right) - 1\right)x_1 - (\kappa - x_1^2 - x_2^2)\sin\left(\frac{\pi}{4}\right)x_2,$$
$$\Delta x_2|_{x \in s} = (\kappa - x_1^2 - x_2^2)\sin\left(\frac{\pi}{4}\right)x_1 + \left((\kappa - x_1^2 - x_2^2)\cos\left(\frac{\pi}{4}\right) - 1\right)x_2,$$
$$(2.1.36)$$

where s is given by the equation $x_2 = x_1$, $x_1 > 0$, the origin is a stable focus. Indeed, using polar coordinates, denote by $r(\phi, r_0)$ the solution of (2.1.36) starting at the angle $\phi = \frac{\pi}{4}$. We can define that $r(\frac{\pi}{4} + 2\pi n, r_0) = (\kappa - r^2(\frac{\pi}{4} + 2\pi(n - 1), r_0))\exp(-\frac{\pi}{4})$. From the last expression, it is easily seen that the sequence $r_n = r(\frac{\pi}{4} + 2\pi n, r_0)$ is monotonically decreasing and there exists a limit of r_n. Assume that $r_n \to \sigma \neq 0$. Then, it implies that there exists a periodic solution of (2.1.36) and $\sigma = (\kappa - \sigma^2)\exp(-\frac{7\pi}{4})\sigma$ which is a contradiction. Thus, $\sigma = 0$. Consequently, the origin is a stable focus of (2.1.36), and by Theorem 2.1.4 the system (2.1.35) has a limit cycle with period $\approx \frac{7\pi}{4}$ if $\mu > 0$ is sufficiently small.

2.2 3D Discontinuous Cycles

We consider three-dimensional discontinuous dynamical systems with nonfixed moments of impacts. Existence of the center manifold is proved for the system. The result is applied for the extension of the planar Hopf bifurcation theorem in Sect. 2.1. Illustrative examples are constructed for the theory.

2.2.1 Introduction

Dynamical systems are used to describe the real-world motions using differential (continuous time) or difference (discrete time) equations. In the last several decades, the need for discontinuous dynamical systems has been increased because they, often, describe the model better when the discontinuous and continuous motions are mingled. This need has made scientists to improve and develop the theory of these systems. Many new results have raised. One must mention that namely systems with not prescribed time of discontinuities were, apparently, introduced for investigation of the real world firstly [73, 185], and this fact emphasizes very much the practical sense of the theory. The problem is one of the most difficult and interesting subjects of investigations [107, 117, 157, 158, 165, 177, 208]. It was emphasized in early stage of theory's development [176].

In the previous section, the Hopf bifurcation for the planar discontinuous dynamical system has been studied. Here, we extend this result to three-dimensional space based on the center manifold. The advantage is that we use the method of B-equivalence [1, 5] as well as the results of timescales which are developed in [38].

This section is organized as follows. In the next section, we start to analyze the nonperturbed system. Section 2.2.3 describes the perturbed system. The center manifold is given in Sect. 2.2.4. In Sect. 2.2.5, the bifurcation of periodic solutions is studied. Section 2.2.6 is devoted to examples in order to illustrate the theory.

2.2.2 The Nonperturbed System

Let $\mathbb{N}$, $\mathbb{R}$ be the sets of all natural and real numbers, respectively, and $\mathbb{R}^2$ be a real euclidean space. Denote by $\langle x, y \rangle$ the dot product of vectors $x, y \in \mathbb{R}^2$. Let $\|x\| = \langle x, x \rangle^{1/2}$ be the norm of a vector $x \in \mathbb{R}^2$, $\mathbb{R}^{2 \times 2}$ be the set of real-valued constant 2×2 matrices, and $I \in \mathbb{R}^{2 \times 2}$ be the identity matrix. We shall consider in $\mathbb{R}^3$ the following dynamical system:

$$\frac{dx}{dt} = Ax,$$

$$\frac{dz}{dt} = \hat{b}z, \quad (x, z) \notin \Gamma_0,$$

$$\Delta x \mid_{(x,z)\in\Gamma_0} = B_0 x,$$

$$\Delta z \mid_{(x,z)\in\Gamma_0} = c_0 z,$$

(2.2.37)

where $A, B_0 \in \mathbb{R}^{2\times2}, \hat{b}, c_0 \in \mathbb{R}, \Gamma_0$ is a subset of $\mathbb{R}^3$ and will be described below. The phase point of (2.2.37) moves between two consecutive intersections with the set Γ_0 along one of the trajectories of the system $x' = Ax, z' = \hat{b}z$. When the solution meets the set Γ_0 at the moment τ, the point $x(t)$ has a jump $\Delta x \mid_\tau := x(\tau+) - x(\tau)$ and the point $z(t)$ has a jump $\Delta z \mid_\tau := z(\tau+) - z(\tau)$. Thus, we suppose that the solutions are left continuous functions.

From now on, G denotes a neighborhood of the origin.

The following assumptions will be needed:

(C1) $\Gamma_0 = \bigcup_{i=1}^{p} \mathscr{P}_i, p \in \mathbb{N}$, where $\mathscr{P}_i = \ell_i \times \mathbb{R}, \ell_i$ are half-lines starting at the origin defined by $\langle a^i, x \rangle = 0$ for $i = 1, \ldots, p$ and $a^i = (a_1^i, a_2^i) \in \mathbb{R}^2$ are constant vectors;

(C2) $A = \begin{bmatrix} \alpha & -\beta \\ \beta & \alpha \end{bmatrix}$, where $\beta \neq 0$;

(C3) There exists a regular matrix $Q \in \mathbb{R}^{2\times2}$ and nonnegative real numbers k and θ such that

$$B_0 = kQ \begin{bmatrix} \cos\theta & -\sin\theta \\ \sin\theta & \cos\theta \end{bmatrix} Q^{-1} - \begin{bmatrix} 1 & 0 \\ 0 & 1 \end{bmatrix}.$$

For the sake of brevity, in what follows, every angle for a point or a line is considered with respect to the half-line of the first coordinate axis in x-plane. Denote $\ell_i' = (I + B_0)\ell_i, i = 1, \ldots, p$. Let γ_i and ζ_i be the angles of ℓ_i and ℓ_i' for $i = 1, \ldots, p$, respectively, and $B_0 = \begin{bmatrix} b_{11} & b_{12} \\ b_{21} & b_{22} \end{bmatrix}$;

(C4) $0 < \gamma_1 < \zeta_1 < \gamma_2 < \cdots < \gamma_p < \zeta_p < 2\pi$, and $(b_{11} + 1)\cos\gamma_i + b_{12}\sin\gamma_i \neq 0$ for $i = 1, \ldots, p$.

In Fig. 2.5, the discontinuity set and a trajectory of the system (2.2.37) are shown. The planes $\mathscr{P}_i$ form the set Γ_0, and each $\mathscr{P}_i'$ is the image of $\mathscr{P}_i$ under the transformation $(I + B)x$.

The system (2.2.37) is said to be a D_0-$system$ if conditions (C1)–(C4) hold. It is easy to see that the origin is a unique singular point of D_0-$system$ and (2.2.37) is not linear.

Let us subject (2.2.37) to the transformation $x_1 = r\cos\phi, x_2 = r\sin\phi, z = z$ and exclude the time variable t. The solution $(r(\phi, r_0, z_0), z(\phi, r_0, z_0))$ which starts at the point $(0, r_0, z_0)$ satisfies the following system in cylindrical coordinates:

Fig. 2.5 The discontinuity
set and a trajectory of
(2.2.37)

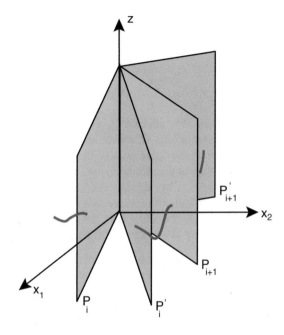

$$\frac{dr}{d\phi} = \lambda r,$$
$$\frac{dz}{d\phi} = bz, \quad \phi \neq \gamma_i \ (\text{mod } 2\pi),$$
$$\Delta r \mid_{\phi = \gamma_i \ (\text{mod } 2\pi)} = k_i r,$$
$$\Delta z \mid_{\phi = \gamma_i \ (\text{mod } 2\pi)} = c_0 z,$$

$\qquad\qquad\qquad\qquad\qquad\qquad\qquad\qquad\qquad\qquad$ (2.2.38)

where $\lambda = \alpha/\beta$, $b = \hat{b}/\beta$, and the variable ϕ is ranged over the timescale

$$\mathbb{R}_\phi = \mathbb{R} \setminus \bigcup_{i=-\infty}^{\infty} \bigcup_{j=1}^{p} (2\pi i + \gamma_j, 2\pi i + \zeta_j]$$

and

$$k_i = \left[((b_{11} + 1) \cos \gamma_i + b_{12} \sin \gamma_i)^2 + (b_{21} \cos \gamma_i + (b_{22} + 1) \sin \gamma_i)^2 \right]^{1/2} - 1.$$

Equation (2.2.38) is 2π-periodic, so, in what follows, we shall consider just the
section $[0, 2\pi]$. That is, the system

$$\frac{dr}{d\phi} = \lambda r,$$

$$\frac{dz}{d\phi} = bz, \quad \phi \neq \gamma_i,$$

$$\Delta r \mid_{\phi=\gamma_i} = k_i r,$$

$$\Delta z \mid_{\phi=\gamma_i} = c_0 z, \qquad (2.2.39)$$

is provided for discussion, where $\phi \in [0, 2\pi]_\phi = [0, 2\pi] \setminus \cup_{i=1}^{p}(\gamma_i, \zeta_i]$. System (2.2.39) is a sample of timescale differential equation. Let us use the ψ-*substitution*, $\varphi = \psi(\phi) = \phi - \sum_{0<\gamma_j<\phi} \theta_j, \theta_j = \zeta_j - \gamma_j$, which was introduced and developed in [6, 38]. The range of this new variable is $[0, 2\pi - \sum_{i=1}^{p} \theta_i]$.

It is easy to check that upon ψ-substitution (2.2.39) reduces to the following impulsive equations:

$$\frac{dr}{d\varphi} = \lambda r,$$

$$\frac{dz}{d\varphi} = bz, \quad \varphi \neq \varphi_i,$$

$$\Delta r \mid_{\varphi=\varphi_i} = k_i r,$$

$$\Delta z \mid_{\varphi=\varphi_i} = c_0 z, \qquad (2.2.40)$$

where $\varphi_i = \psi(\gamma_i)$. Solving (2.2.40) as an impulsive system [156, 215] and using ψ-substitution, one can obtain that a solution of (2.2.39) is of the form

$$r(\phi) = \exp\left(\lambda\left(\phi - \sum_{0<\gamma_i<\phi} \theta_i\right)\right)\left[\prod_{0<\gamma_i<\phi}(1 + k_i)\right] r_0, \qquad (2.2.41)$$

$$z(\phi) = \exp\left(b\left(\phi - \sum_{0<\gamma_i<\phi} \theta_i\right)\right)\left[\prod_{0<\gamma_i<\phi}(1 + c_0)\right] z_0, \qquad (2.2.42)$$

for $\phi \in [0, 2\pi]_\phi$. Denote

$$q_1 = \exp\left(\lambda\left(2\pi - \sum_{i=1}^{p} \theta_i\right)\right)\prod_{i=1}^{p}(1 + k_i), \qquad (2.2.43)$$

$$q_2 = \exp\left(b\left(2\pi - \sum_{i=1}^{p} \theta_i\right)\right)\prod_{i=1}^{p}(1 + c_0). \qquad (2.2.44)$$

Depending on q_1 and q_2, we may see that the following lemmas are valid.

Lemma 2.2.1 *Assume that* $q_1 = 1$. *Then, if*

(i) $q_2 = 1$ *then all solutions are periodic with period* $T = \left(2\pi - \sum_{i=1}^{p} \theta_i\right)\beta^{-1}$;

(ii) $q_2 = -1$ *then a solution that starts to its motion on* x_1x_2-*plane is* T-*periodic and all other solutions are* $2T$-*periodic;*

(iii) $| q_2 |> 1$ *then a solution that starts to its motion on $x_1 x_2$-plane is T-periodic and all other solutions lie on the surface of a cylinder and they move away the origin (i.e., zero solution is unstable);*

(iv) $| q_2 |< 1$ *then a solution that starts to its motion on $x_1 x_2$-plane is T-periodic and all other solutions lie on the surface of a cylinder and they move toward the $x_1 x_2$-plane (i.e., zero solution is stable).*

Lemma 2.2.2 *Assume that $q_1 < 1$. Then, if*

(i) $| q_2 |< 1$ *all solutions will spiral toward the origin, i.e., origin is an asymptotically stable fixed point;*

(ii) $| q_2 |> 1$ *a solution that starts to its motion on x-plane spirals toward the origin and a solution that starts to its motion on z-axis will move away from the origin. In this case the origin is half stable (or conditionally stable);*

(iii) $q_2 = 1 (q_2 = -1)$ *then a solution that starts to its motion on z-axis is periodic with period $T (2T)$ and all other solutions will approach to z-axis.*

Lemma 2.2.3 *Assume that $q_1 > 1$. Then, if*

(i) $| q_2 |< 1$ *then origin is a stable focus;*

(ii) $| q_2 |> 1$ *then origin is an unstable focus;*

(iii) $q_2 = 1 (q_2 = -1)$ *then a solution that starts to its motion on z-axis is periodic with period $T (2T)$ and all other solutions will approach to z-axis.*

We note that when $q_2 = -1$ (this means z may be negative, too), the solutions starting their motion out of $x_1 x_2$-plane will move above and below the $x_1 x_2$-plane. More explicitly, if a solution starts to its motion above the x-plane, then after the time corresponding to an angle of T, it will be below the x-plane; in the next duration corresponding to an angle T, it will try to move above x-plane; and at the end of that duration, it will be above the x-plane, and so on.

From now on, we assume that $q_1 = 1$ and $| q_2 |< 1$.

2.2.3 The Perturbed System

Consider the system

$$\frac{dx}{dt} = Ax + f(x, z),$$
$$\frac{dz}{dt} = \hat{b}z + g(x, z), \quad (x, z) \notin \Gamma,$$
$$\Delta x \mid_{(x,z)\in\Gamma} = B(x)x,$$
$$\Delta z \mid_{(x,z)\in\Gamma} = c(z)z,$$

(2.2.45)

where the followings are assumed to be true.

(C5) $\Gamma = \bigcup_{i=1}^{p} \mathscr{S}_i$, where $\mathscr{S}_i = s_i \times \mathbb{R}$ and the equation of s_i is given by s_i : $\langle a^i, x \rangle + \tau_i(x) = 0$, for $i = 1, \ldots, p$;
(C6)

$$B(x) = (k + \kappa(x))Q \begin{bmatrix} \cos(\theta + \Theta(x)) & -\sin(\theta + \Theta(x)) \\ \sin(\theta + \Theta(x)) & \cos(\theta + \Theta(x)) \end{bmatrix} Q^{-1} - \begin{bmatrix} 1 & 0 \\ 0 & 1 \end{bmatrix}$$

and $c(z) = c_0 + \tilde{c}(z)$;
(C7) Functions $f, g, \kappa, \tilde{c}$, and Θ are in C^1 and τ_i is in C^2;
(C8) $f(x, z) = \mathscr{O}(\|(x, z)\|)$, $g(x, z) = \mathscr{O}(\|(x, z)\|)$, $\kappa(x) = \mathscr{O}(\|x\|)$, $\Theta(x) = \mathscr{O}(\|x\|)$, $\tilde{c}(z) = \mathscr{O}(z)$, $\tau_i(x) = \mathscr{O}(\|x\|^2)$, $i = 1, \ldots, p$.

Moreover, it is supposed that the matrices A, Q, the vectors $a^i, i = 1, \ldots, p$, constants k, θ are the same as for (2.2.37), i.e.,

(C9) The associated with (2.2.45) is D_0 system.

Remark 2.2.1 Conditions (C5) and (C6) imply that surfaces $\mathscr{S}_i$ do not intersect each other except on z-axis and neither of them intersects itself.

The system (2.2.45) is said to be a *D-system* if the conditions $(C1)$–$(C8)$ hold.
In what follows, we assume without loss of generality that $\gamma_i \neq \frac{\pi}{2}j, j = 1, 2, 3$. Then, one can transform the equation in (C5) to the polar coordinates so that s_i : $a_i^1 r \cos \phi + a_i^2 r \sin \phi + \tau_i(r \cos \phi, r \sin \phi) = 0$ and, hence

$$\phi = \tan^{-1} \left(\tan \gamma_i - \frac{\tau_i(r \cos \phi, r \sin \phi)}{a_i^2 r \cos \phi} \right).$$

Using Taylor expansion gives that the previous equation can be written, for sufficiently small r, as
$$s_i : \phi = \gamma_i + r \Psi_i(r, \phi), i = 1, \ldots, p$$

where functions Ψ_i are 2π-periodic in ϕ, continuously differentiable and $\Psi_i = \mathscr{O}(r)$.
If the phase point $(x_1(t), x_2(t), z(t))$ meets the discontinuity surface $\mathscr{S}_i$ at the angle θ, then after the jump the point $(x_1(\theta+), x_2(\theta+), z(\theta+))$ will belong to the surface $\mathscr{S}_i' = \{(u, v) \in \mathbb{R}^3 : u = (I + B(x))x, v = (1 + c_0)z + c(z), (x, z) \in \mathscr{S}_i\}$. For the remaining part of this section, the following assertion is very important and the proof can be found in [6].

Lemma 2.2.4 *If the conditions (C7) and (C8) are valid then the surface $\mathscr{S}_i'$ is placed between the surfaces $\mathscr{S}_i$ and $\mathscr{S}_{i+1}$ for every i if G is sufficiently small.*

Using the cylindrical coordinates $x_1 = r \cos \phi, x_2 = r \sin \phi, \angle = z$, one can find that the differential part of (2.2.45) has the following form:

$$\frac{dr}{d\phi} = \lambda r + P(r, \phi, z),$$
$$\frac{dz}{d\phi} = bz + Q(r, \phi, z),$$

(2.2.46)

where, as is known [218], the functions $P(r, \phi, z)$ and $Q(r, \phi, z)$ are 2π-periodic in ϕ, continuously differentiable, and $P = \mathcal{O}(r, z)$, $Q = \mathcal{O}(r, z)$. Denote $x^+ = (x_1^+, x_2^+) = (I + B(x))x$, $x^+ = r^+(\cos\phi^+, \sin\phi^+)$, $\tilde{x}^+ = (\tilde{x}_1^+, \tilde{x}_2^+) = (I + B(0))x$, where $x = (x_1, x_2) \in s_i$, $i = 1, \dots, p$. The inequality $\|x^+ - \tilde{x}^+\| \leq \|B(x) - B(0)\| \cdot \|x\|$ implies that $r^+ = (1 + k_i)r + \omega(r, \phi)$. Moreover, using the relation $\frac{x_2^+}{x_1^+}$ and $\frac{\tilde{x}_2^+}{\tilde{x}_1^+}$ and condition (C5), one can conclude that $\phi^+ = \phi + \theta_i + \gamma(r, \phi)$. Functions ω and γ are 2π-periodic in ϕ and $\omega = \mathcal{O}(r)$, $\gamma = \mathcal{O}(r)$. Finally, transformed system (2.2.45) is of the following form:

$$
\begin{aligned}
\frac{dr}{d\phi} &= \lambda r + P(r, \phi, z), \\
\frac{dz}{d\phi} &= bz + Q(r, \phi, z), \quad (r, \phi, z) \notin \Gamma, \\
\Delta r\,|_{(r,\phi)\in s_i} &= k_i r + \omega(r, \phi), \\
\Delta \phi\,|_{(r,\phi)\in s_i} &= \theta_i + \gamma(r, \phi), \\
\Delta z\,|_{(r,\phi)\in s_i} &= c_0 z + \tilde{c}(z).
\end{aligned}
\tag{2.2.47}
$$

Let us introduce the following system besides (2.2.47):

$$
\begin{aligned}
\frac{d\rho}{d\phi} &= \lambda \rho + P(\rho, \phi, z), \\
\frac{dz}{d\phi} &= bz + Q(\rho, \phi, z), \quad \phi \neq \gamma_i, \\
\Delta \rho\,|_{\phi=\gamma_i} &= k_i \rho + W_i^1(\rho, z), \\
\Delta \phi\,|_{\phi=\gamma_i} &= \theta_i, \\
\Delta z\,|_{\phi=\gamma_i} &= c_0 z + W_i^2(\rho, z),
\end{aligned}
\tag{2.2.48}
$$

where all elements, except for $W_i = (W_i^1, W_i^2)$, $i = 1, \dots, p$, are the same as in (2.2.47) and the domain of (2.2.48) is $[0, 2\pi]_\phi$. We shall define the functions W_i below.

Let $(r(\phi, r_0, z_0), z(\phi, r_0, z_0))$ be a solution of (2.2.47) ϕ_i be the angle where the phase point intersects $\mathscr{S}_i$. Denote also by $\chi_i = \phi_i + \theta_i + \gamma(r(\phi_i, r_0, z_0), \phi_i)$ the angle where the phase point has to be after the jump.

Further $(\alpha, \widehat{\beta}]$, $\{\alpha, \beta\} \subset \mathbb{R}$ denotes the oriented interval, that is

$$
(\alpha, \widehat{\beta}] = \begin{cases} (\alpha, \beta] \text{ if } \alpha \leq \beta, \\ (\beta, \alpha] \text{ otherwise.} \end{cases}
$$

Definition 2.2.1 We shall say that systems (2.2.47) and (2.2.48) are B-equivalent in G if for every solution $(r(\phi, r_0, z_0), z(\phi, r_0, z_0))$ of (2.2.47) whose trajectory is in G for all $\phi \in [0, 2\pi]_\phi$ there exists a solution $(\rho(\phi, r_0, z_0), z(\phi, r_0, z_0))$ of (2.2.48) which satisfies the relation

$$
r(\phi, r_0, z_0) = \rho(\phi, r_0, z_0), \quad \phi \in [0, 2\pi]_\phi \setminus \bigcup_{i=1}^{p}\{(\phi_i, \widehat{\gamma_i}] \cup (\zeta_i, \widehat{\chi_i}]\}, \tag{2.2.49}
$$

and, conversely, for every solution $(\rho(\phi, r_0, z_0), z(\phi, r_0, z_0))$ of (2.2.48) whose trajectory is in G, there exists a solution $(r(\phi, r_0, z_0), z(\phi, r_0, z_0))$ of (2.2.47) which satisfies (2.2.49).

Fix $i = 1, \ldots, p$. Let $(r_1(\phi), z_1(\phi))$, $(r_1(\gamma_i), z_1(\gamma_i)) = (\rho, z)$, be a solution of

$$\frac{dr}{d\phi} = \lambda r + P(r, \phi, z),$$
$$\frac{dz}{d\phi} = bz + Q(r, \phi, z),$$

(2.2.50)

and let $\phi = \eta_i$ be the meeting angle of the solution with $\mathscr{P}_i$. Then

$$r_1(\eta_i) = e^{\lambda(\eta_i - \gamma_i)} \rho + \int_{\gamma_i}^{\eta_i} e^{\lambda(\eta_i - s)} P(r_1(s), s, z_1(s)) ds,$$

$$z_1(\eta_i) = e^{b(\eta_i - \gamma_i)} z + \int_{\gamma_i}^{\eta_i} e^{b(\eta_i - s)} Q(r_1(s), s, z_1(s)) ds.$$

Let $\eta_i' = \eta_i + \theta_i + \gamma(r_1(\eta_i), \eta_i)$ and $(\rho', z') = ((1 + k_i)r_1(\eta_i) + \omega(r_1(\eta_i), \eta_i), (1 + c_0)z_1(\eta_i) + c(z_1(\eta_i)))$. Let $(r_2(\phi), z_2(\phi))$, $(r_2(\eta_i'), z_2(\eta_i')) = (\rho', z')$, be a solution of (2.2.50). Then,

$$r_2(\zeta_i) = e^{\lambda(\zeta_i - \eta_i')} \rho' + \int_{\eta_i'}^{\zeta_i} e^{\lambda(\zeta_i - s)} P(r_2(s), s, z_2(s)) ds,$$

$$z_2(\zeta_i) = e^{b(\zeta_i - \eta_i')} z' + \int_{\eta_i'}^{\zeta_i} e^{b(\zeta_i - s)} Q(r_2(s), s, z_2(s)) ds.$$

We define that

$$W_i^1(\rho, z) = r_2(\zeta_i) - (1 + k_i)\rho$$
$$= e^{\lambda(\zeta_i - \eta_i')} \left[(1 + k_i) \left(e^{\lambda(\eta_i - \gamma_i)} \rho + \int_{\gamma_i}^{\eta_i} e^{\lambda(\eta_i - s)} P(r_1(s), s, z_1(s)) ds \right) \right.$$
$$\left. + \omega(r_1(\eta_i), \eta_i) \right] + \int_{\eta_i'}^{\zeta_i} e^{\lambda(\zeta_i - s)} P(r_1(s), s, z_1(s)) ds - (1 + k_i)\rho,$$

or, if simplified

$$W_i^1(\rho, z) = (1 + k_i)(e^{-\lambda\gamma(r_1(\eta_i), \eta_i)} - 1)\rho$$
$$+ (1 + k_i) \int_{\gamma_i}^{\eta_i} e^{\lambda(\zeta_i - \theta_i - s - \gamma(r_1(\eta_i), \eta_i))} P(r_1(s), s, z_1(s)) ds$$
$$+ \int_{\eta_i'}^{\zeta_i} e^{\lambda(\zeta_i - s)} P(r_2(s), s, z_2(s)) ds + e^{\lambda(\zeta_i - \eta_i')} \omega(r_1(\eta_i), \eta_i). \quad (2.2.51)$$

We, similarly, define

$$
\begin{aligned}
W_i^2(\rho, z) &= z_2(\zeta_i) - (1 + c_0)z \\
&= e^{b(\zeta_i - \eta_i')}\left[(1 + c_0)\left(e^{b(\eta_i - \gamma_i)}z + \int_{\gamma_i}^{\eta_i} e^{b(\eta_i - s)}Q(r_1(s), s, z_1(s))ds\right)\right. \\
&\quad \left. + \tilde{c}(z_1(\eta_i))\right] + \int_{\eta_i'}^{\zeta_i} e^{b(\zeta_i - s)}Q(r_1(s), s, z_1(s))ds - (1 + c_0)z,
\end{aligned}
$$

or,

$$
\begin{aligned}
W_i^2(\rho, z) &= (1 + k_i)(e^{-b\gamma(r_1(\eta_i), \eta_i)} - 1)z \\
&\quad + (1 + c_0)\int_{\gamma_i}^{\eta_i} e^{(\zeta_i - \theta_i - s - \gamma(r_1(\eta_i), \eta_i))}Q(r_1(s), s, z_1(s))ds \\
&\quad + \int_{\eta_i'}^{\zeta_i} e^{b(\zeta_i - s)}Q(r_2(s), s, z_2(s))ds + e^{b(\zeta_i - \eta_i')}\tilde{c}(z_1(\eta_i)). \quad (2.2.52)
\end{aligned}
$$

We note that there exists a Lipschitz constant ℓ and a bounded function $m(\ell)$ such that

$$
\|W_i^j(\rho_1, z_1) - W_i^j(\rho_2, z_2)\| \le m(\ell)\ell(\|\rho_1 - \rho_2\| + \|z_1 - z_2\|), \quad (2.2.53)
$$

for all $\rho_1, \rho_2, z_1, z_2 \in \mathbb{R}$, $j = 1, 2$. For detailed proof and explanation about (2.2.53), we refer to [1, 6, 38].

2.2.4 Center Manifold

Now, using ψ-substitution (2.2.48) reduces to the following system:

$$
\begin{aligned}
\frac{d\rho}{d\varphi} &= \lambda\rho + F(\rho, \varphi, z), \\
\frac{dz}{d\varphi} &= bz + G(\rho, \varphi, z), \quad \varphi \ne \varphi_i, \\
\Delta\rho\,|_{\varphi = \varphi_i} &= k_i\rho + W_i^1(\rho, z), \\
\Delta z\,|_{\varphi = \varphi_i} &= c_0 z + W_i^2(\rho, z),
\end{aligned} \quad (2.2.54)
$$

where $\varphi = \psi(\phi)$, $\varphi_i = \psi(\gamma_i)$, $F(\rho, \varphi, z) = P(\rho, \psi^{-1}(\varphi), z)$, and $G(\rho, \varphi, z) = Q(\rho, \psi^{-1}(\varphi), z)$. Functions F and G are T-periodic in φ, with $T = \psi(2\pi)$, and satisfy

$$
\|F(\rho, \varphi, z) - F(\rho', \varphi, z')\| \le k(\epsilon)(\|\rho - \rho'\| + \|z - z'\|), \quad (2.2.55)
$$
$$
\|G(\rho, \varphi, z) - G(\rho', \varphi, z')\| \le k(\epsilon)(\|\rho - \rho'\| + \|z - z'\|). \quad (2.2.56)
$$

Following the methods given in [8], one can see that system (2.2.54) has two integral manifolds whose equations are given by:

$$\Phi_+(\varphi, \rho) = \int_{-\infty}^{\varphi} \pi_+(\varphi, s) G(\rho(s, \varphi, \rho), s, z(s, \varphi, \rho)) ds$$
$$+ \sum_{\varphi_i < \varphi} \pi_+(\varphi, \varphi_i^+) W_i^2(\rho(\varphi_i^+, \varphi, \rho), z(\varphi_i^+, \varphi, \rho)), \quad (2.2.57)$$

and

$$\Phi_-(\varphi, z) = -\int_{\varphi}^{\infty} \pi_-(\varphi, s) F(\rho(s, \varphi, z), s, z(s, \varphi, z)) ds$$
$$+ \sum_{\varphi_i < \varphi} \pi_-(\varphi, \varphi_i^+) W_i^1(\rho(\varphi_i^+, \varphi, z), z(\varphi_i^+, \varphi, z)), \quad (2.2.58)$$

where

$$\pi_+(\varphi, s) = e^{b(\varphi - s)} \prod_{s \le \varphi_j < \varphi} (1 + c_0)$$

and

$$\pi_-(\varphi, s) = e^{\lambda(\varphi - s)} \prod_{s \le \varphi_j < \varphi} (1 + k_j).$$

In (2.2.57), the pair $(\rho(s, \varphi, \rho), z(s, \varphi, \rho))$ denotes a solution of (2.2.54) satisfying $\rho(\varphi, \varphi, \rho) = \rho$. Similarly, $(\rho(s, \varphi, z), z(s, \varphi, z))$, in (2.2.58), is the solution of (2.2.54) with $z(\varphi, \varphi, z) = z$.

In [8], it was shown that there exist constants K_+, M_+, σ_+ such that Φ_+ satisfies:

$$\Phi_+(\varphi, 0) = 0, \quad (2.2.59)$$
$$\|\Phi_+(\varphi, \rho_1) - \Phi_+(\varphi, \rho_2)\| \le K_+ \ell \|\rho_1 - \rho_2\|, \quad (2.2.60)$$

for all ρ_1, ρ_2 such that a solution $w(\varphi) = (\rho(\varphi), z(\varphi))$ of (2.2.54) with $w(\varphi_0) = (\rho_0, \Phi_+(\varphi_0, \rho_0))$, $\rho_0 \ge 0$, is defined on $\mathbb{R}$ and satisfies

$$\|w(\varphi)\| \le M_+ \rho_0 e^{-\sigma_+(\varphi - \varphi_0)}, \quad \varphi \ge \varphi_0. \quad (2.2.61)$$

Similarly, it was shown that there exist constants K_-, M_-, σ_- such that Φ_- satisfies:

$$\Phi_-(\varphi, 0) = 0, \quad (2.2.62)$$
$$\|\Phi_-(\varphi, z_1) - \Phi_-(\varphi, z_2)\| \le K_- \ell \|z_1 - z_2\|, \quad (2.2.63)$$

for all z_1, z_2 such that a solution $w(\varphi) = (\rho(\varphi), z(\varphi))$ of (2.2.54) with $w(\varphi_0) = (\Phi_-(\varphi_0, z_0), z_0)$, $z_0 \in \mathbb{R}$, is defined on $\mathbb{R}$ and satisfies

$$\|w(\varphi)\| \le M_- \|z_0\| e^{-\sigma_-(\varphi-\varphi_0)}, \quad \varphi \le \varphi_0. \tag{2.2.64}$$

Set $S_+ = \{(\rho, \varphi, z) : z = \Phi_+(\varphi, \rho)\}$ and $S_- = \{(\rho, \varphi, z) : \rho = \Phi_-(\varphi, z)\}$. Here, S_+ is called the *center manifold* and S_- is called the *stable manifold*. A sketch of an arbitrary center manifold is shown in Fig. 2.6.

The analogues of the following two Lemma's together with their proofs can be found in [8].

Lemma 2.2.5 *If the Lipschitz constant ℓ is sufficiently small, then for every solution $w(\varphi) = (\rho(\varphi), z(\varphi))$ of (2.2.54) there exists a solution $\mu(\varphi) = (u(\varphi), v(\varphi))$ on the center manifold, S_+, such that*

$$\begin{aligned}
\|\rho(\varphi) - u(\varphi)\| &\le 2M_+ \|\rho(\varphi_0) - u(\varphi_0)\| e^{-\sigma_+(\varphi-\varphi_0)}, \\
\|z(\varphi) - v(\varphi)\| &\le M_+ \|z(\varphi_0) - v(\varphi_0)\| e^{-\sigma_+(\varphi-\varphi_0)}, \quad \varphi \ge \varphi_0,
\end{aligned} \tag{2.2.65}$$

where M_+ and σ_+ are the constants used in (2.2.61).

Lemma 2.2.6 *For sufficiently small Lipschitz constant ℓ the surface S_+ is stable in large.*

On the local center manifold S_+, the first coordinate of the solutions of (2.2.54) satisfies the following system:

$$\begin{aligned}
\frac{d\rho}{d\varphi} &= \lambda\rho + F(\rho, \varphi, \Phi_+(\varphi, \rho)), \quad \varphi \ne \varphi_i, \\
\Delta\rho \mid_{\varphi=\varphi_i} &= k_i \rho + W_i^1(\rho, \Phi_+(\varphi, \rho)).
\end{aligned} \tag{2.2.66}$$

Fig. 2.6 The center manifold

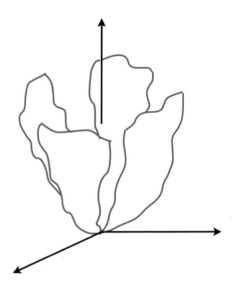

Now, it is time to consider the reduction principle for which we need, together with the ones imposed before, the condition:

(C10) Functions $f(x, z)$ and $g(x, z)$ are continuously differentiable in x, z for all x, z, and

$$\frac{\partial f(0, 0)}{\partial x_j} = 0, \quad \frac{\partial f(0, 0)}{\partial z} = 0, \quad \frac{\partial g(0, 0)}{\partial x_j} = 0, \quad \frac{\partial g(0, 0)}{\partial z} = 0,$$

for $j = 1, 2$ where $x = (x_1, x_2)$.

Theorem 2.2.1 *Assume that conditions (C1)–(C10) are fulfilled. Then the trivial solution of (2.2.54) is stable, asymptotically stable or unstable if the trivial solution of (2.2.66) is stable, asymptotically stable or unstable, respectively.*

Using inverse of ψ-substitution and B-equivalence, one can see that the following theorem holds:

Theorem 2.2.2 *Assume that conditions (C1)–(C10) are fulfilled. Then the trivial solution of (2.2.45) is stable, asymptotically stable or unstable if the trivial solution of (2.2.66) is stable, asymptotically stable or unstable, respectively.*

2.2.5 Bifurcation of Periodic Solutions

This section is devoted to the bifurcation theorem of a periodic solution for the discontinuous dynamical system. Let us consider the system,

$$\begin{aligned}
\frac{dx}{dt} &= Ax + f(x, z) + \mu \tilde{f}(x, z, \mu), \\
\frac{dz}{dt} &= \hat{b}z + g(x, z) + \mu \tilde{g}(x, z, \mu), \quad (x, z) \notin \Gamma(\mu), \\
\Delta x \mid_{(x,z) \in \Gamma(\mu)} &= B(x, \mu)x, \\
\Delta z \mid_{(x,z) \in \Gamma(\mu)} &= c(z, \mu)z.
\end{aligned} \qquad (2.2.67)$$

Assume that the following conditions are satisfied:

(A1) The set $\Gamma(\mu) = \bigcup_{i=1}^{p} \mathscr{S}_i(\mu)$, where $\mathscr{S}_i(\mu) = s_i(\mu) \times \mathbb{R}$ and the equation of $s_i(\mu)$ is given by $s_i(\mu) : \langle a^i, x \rangle + \tau_i(x) + \mu v(x, \mu) = 0$, for $i = 1, \ldots, p$;

(A2) There exists a matrix $Q(\mu) \in \mathbb{R}^{2 \times 2}$, $Q(0) = Q$, analytic in $(-\mu_0, \mu_0)$, and real numbers γ, χ such that $Q^{-1}(\mu)B(x, \mu)Q(\mu) =$

$$(k + \mu\gamma + \kappa(x)) \begin{bmatrix} \cos(\theta + \mu\chi + \Theta(x)) & -\sin(\theta + \mu\chi + \Theta(x)) \\ \sin(\theta + \mu\chi + \Theta(x)) & \cos(\theta + \mu\chi + \Theta(x)) \end{bmatrix} - \begin{bmatrix} 1 & 0 \\ 0 & 1 \end{bmatrix}$$

and $c(z, \mu) = c_0 + \tilde{c}(z) + \mu d(z, \mu)$;

(A3) Associated with (2.2.67) systems

$$
\begin{aligned}
\frac{dx}{dt} &= Ax, \\
\frac{dz}{dt} &= \hat{b}z, \quad (x, z) \notin \Gamma_0, \\
\Delta x \mid_{(x,z)\in\Gamma_0} &= B_0 x, \\
\Delta z \mid_{(x,z)\in\Gamma_0} &= c_0 z.
\end{aligned}
\tag{2.2.68}
$$

and

$$
\begin{aligned}
\frac{dx}{dt} &= Ax + f(x, z), \\
\frac{dz}{dt} &= \hat{b}z + g(x, z), \quad (x, z) \notin \Gamma(0), \\
\Delta x \mid_{(x,z)\in\Gamma(0)} &= B(x, 0)x, \\
\Delta z \mid_{(x,z)\in\Gamma(0)} &= c(z, 0)z.
\end{aligned}
\tag{2.2.69}
$$

are D_0-system and D-system, respectively;
(A4) Functions $\tilde{f}, v : G \times (-\mu_0, \mu_0) \to \mathbb{R}^2$ are analytic in $x, z,$ and μ;
(A5) $\tilde{f}(0, 0, \mu) = 0, v(0, \mu) = 0,$ uniformly for $\mu \in (-\mu_0, \mu_0)$.

Using polar coordinates, one can write system (2.2.67) in the following form:

$$
\begin{aligned}
\frac{dr}{d\phi} &= \lambda(\mu)r + P(r, \phi, z, \mu), \\
\frac{dz}{d\phi} &= b(\mu)z + Q(r, \phi, z, \mu), \quad (r, \phi, z) \notin \Gamma(\mu), \\
\Delta r \mid_{(r,\phi)\in\ell_i(\mu)} &= k_i(\mu)r + \omega(r, \phi, \mu), \\
\Delta \phi \mid_{(r,\phi)\in\ell_i(\mu)} &= \theta_i(\mu) + \gamma(r, \phi, \mu), \\
\Delta z \mid_{(r,\phi)\in\ell_i(\mu)} &= c_0(\mu)z + \tilde{c}(z, \mu).
\end{aligned}
\tag{2.2.70}
$$

Let the system

$$
\begin{aligned}
\frac{d\rho}{d\phi} &= \lambda(\mu)\rho + P(\rho, \phi, z, \mu), \\
\frac{dz}{d\phi} &= b(\mu)z + Q(\rho, \phi, z, \mu), \quad \phi \neq \gamma_i(\mu), \\
\Delta \rho \mid_{\phi=\gamma_i(\mu)} &= k_i(\mu)\rho + W_i^1(\rho, z, \mu), \\
\Delta \phi \mid_{\phi=\gamma_i(\mu)} &= \theta_i(\mu), \\
\Delta z \mid_{\phi=\gamma_i(\mu)} &= c_0(\mu)z + W_i^2(\rho, z, \mu),
\end{aligned}
\tag{2.2.71}
$$

where $\gamma_i(\mu), i = 1, \ldots, p$, are angles of m_i, be B-equivalent to (2.2.70). The functions $W_i^1(\rho, z, \mu)$ and $W_i^2(\rho, z, \mu)$ can be defined in the same manner as in (2.2.51) and (2.2.52), respectively. Applying ψ-substitution to (2.2.71), we get

$$\frac{d\rho}{d\varphi} = \lambda(\mu)\rho + F(\rho, \varphi, z, \mu),$$

$$\frac{dz}{d\varphi} = b(\mu)z + G(\rho, \varphi, z, \mu), \quad \varphi \neq \varphi_i(\mu), \tag{2.2.72}$$

$$\Delta\rho \mid_{\varphi=\varphi_i(\mu)} = k_i(\mu)\rho + W_i^1(\rho, z, \mu),$$

$$\Delta z \mid_{\varphi=\varphi_i(\mu)} = c_0(\mu)z + W_i^2(\rho, z, \mu).$$

Following the methods, as we did to obtain (2.2.57) and (2.2.58), one can see that system (2.2.72) has two integral manifolds whose equations are given by:

$$\Phi_+(\varphi, \rho, \mu) = \int_{-\infty}^{\varphi} \pi_+(\varphi, s, \mu)G(\rho(s, \varphi, \rho, \mu), s, z(s, \varphi, \rho, \mu), \mu)ds$$

$$+ \sum_{\varphi_i(\mu)<\varphi} \pi_+(\varphi, \varphi_i^+, \mu)W_i^2(\rho(\varphi_i^+, \varphi, \rho, \mu), z(\varphi_i^+, \varphi, \rho, \mu)), \tag{2.2.73}$$

and

$$\Phi_-(\varphi, z, \mu) = -\int_{\varphi}^{\infty} \pi_-(\varphi, s, \mu)F(\rho(s, \varphi, z, \mu), s, z(s, \varphi, z, \mu), \mu)ds$$

$$+ \sum_{\varphi_i(\mu)<\varphi} \pi_-(\varphi, \varphi_i^+, \mu)W_i^1(\rho(\varphi_i^+, \varphi, z, \mu), z(\varphi_i^+, \varphi, z, \mu)), \tag{2.2.74}$$

where

$$\pi_+(\varphi, s, \mu) = e^{b(\varphi-s)} \prod_{s \leq \varphi_j(\mu)<\varphi} (1 + c_0(\mu)),$$

and

$$\pi_-(\varphi, s, \mu) = e^{\lambda(\varphi-s)} \prod_{s \leq \varphi_j(\mu)<\varphi} (1 + k_j(\mu)).$$

In (2.2.73), the pair $(\rho(s, \varphi, \rho, \mu), z(s, \varphi, \rho, \mu))$ denotes a solution of (2.2.72) satisfying $\rho(\varphi, \varphi, \rho, \mu) = \rho$. Similarly, $(\rho(s, \varphi, z, \mu), z(s, \varphi, z, \mu))$, in (2.2.74), is a solution of (2.2.72) with $z(\varphi, \varphi, z, \mu) = z$. Set $S_+(\mu) = \{(\rho, \varphi, z) : z = \Phi_+(\varphi, \rho, \mu)\}$ and $S_-(\mu) = \{(\rho, \varphi, z) : \rho = \Phi_-(\varphi, z, \mu)\}$.

On the local center manifold, $S_+(\mu)$, the first coordinate of the solutions of (2.2.72) satisfies the following system:

$$\frac{d\rho}{d\varphi} = \lambda(\mu)\rho + F(\rho, \varphi, \Phi_+(\varphi, \rho, \mu)), \quad \varphi \neq \varphi_i(\mu), \tag{2.2.75}$$

$$\Delta\rho \mid_{\varphi=\varphi_i(\mu)} = k_i(\mu)\rho + W_i^1(\rho, \Phi_+(\varphi, \rho, \mu)).$$

Similar to (2.2.43) and (2.2.44), one can define the functions

$$q_1(\mu) = \exp\left(\lambda(\mu)\left(2\pi - \sum_{i=1}^{p}\theta_i(\mu)\right)\right)\prod_{i=1}^{p}(1 + k_i(\mu)), \qquad (2.2.76)$$

and

$$q_2(\mu) = \exp\left(b(\mu)\left(2\pi - \sum_{i=1}^{p}\theta_i(\mu)\right)\right)\prod_{i=1}^{p}(1 + c_0(\mu)). \qquad (2.2.77)$$

System (2.2.75) is the system studied in [6], and there it was shown that this system, for sufficiently small μ, has a periodic solution with period T. Here, we will show that if the first coordinate of a solution of (2.2.72) is T-periodic, then so is the second coordinate.

Now, since

$$\pi_+(\varphi + T, s + T, \mu) = \pi_+(\varphi, s, \mu),$$

$$\rho(s + T, \varphi + T, \rho, \mu) = \rho(s, \varphi, \rho, \mu),$$

$$z(s + T, \varphi + T, \rho, \mu) = z(s, \varphi, \rho, \mu),$$

and G is T-periodic in φ, we have,

$$\begin{aligned}
&\Phi_+(\varphi + T, \rho, \mu) \\
&= \int_{-\infty}^{\varphi+T} \pi_+(\varphi + T, s, \mu)G(\rho(s, \varphi + T, \rho, \mu), s, z(s, \varphi + T, \rho, \mu), \mu)ds \\
&\quad + \sum_{\varphi_i(\mu) < \varphi+T} \pi_+(\varphi + T, \varphi_i^+, \mu)W_i^2(\rho(\varphi_i^+, \varphi + T, \rho, \mu), z(\varphi_i^+, \varphi + T, \rho, \mu)) \\
&= \int_{-\infty}^{\varphi} \pi_+(\varphi, t, \mu)G(\rho(t, \varphi, \rho, \mu), t, z(t, \varphi, \rho, \mu), \mu)dt \\
&\quad + \sum_{\bar{\varphi}_i(\mu) < \varphi} \pi_+(\varphi, \bar{\varphi}_i^+, \mu)W_i^2(\rho(\bar{\varphi}_i^+, \varphi, \rho, \mu), z(\bar{\varphi}_i^+, \varphi, \rho, \mu)) \\
&= \Phi_+(\varphi, \rho, \mu)
\end{aligned}$$

Then, we have the following theorem which, in case of two dimension, was shown in Sect. 2.1.

Theorem 2.2.3 *Assume that $q_1(0) = 1$, $q_1'(0) \neq 0$, $\mid q_2(0) \mid < 1$, and the origin is a focus for (2.2.69). Then, for sufficiently small r_0 and z_0, there exists a function $\mu = \delta(r_0, z_0)$ such that the solution $(r(\phi, \delta(r_0, z_0)), z(\phi, \delta(r_0, z_0)))$ of (2.2.70), with the initial condition $r(0, \delta(r_0, z_0)) = r_0$, $z(0, \delta(r_0, z_0)) = z_0$, is periodic with a period, $T' = \left(2\pi - \sum_{i=1}^{p}\theta_i\right)\beta^{-1} + o(|\mu|)$.*

2.2.6 Examples

Example 2.2.1 Consider the following dynamical system:

$$
\begin{aligned}
x_1' &= (0.1 - \mu)x_1 - 20x_2 + 2x_1x_2, \\
x_2' &= 20x_1 + (0.1 - \mu)x_2 + 3x_1^2 z, \\
z' &= (-0.3 + \mu)z + \mu^2 x_1 z, \quad (x_1, x_2, z) \notin \mathscr{S}, \\
\Delta x_1 \mid_{(x_1,x_2,z) \in \mathscr{S}} &= \left((\kappa_1 + \mu^3) \cos(\tfrac{\pi}{3}) - 1 \right) x_1 - (\kappa_1 + \mu^3) \sin(\tfrac{\pi}{3})x_2, \\
\Delta x_2 \mid_{(x_1,x_2,z) \in \mathscr{S}} &= (\kappa_1 + \mu^3) \sin(\tfrac{\pi}{3})x_1 + \left((\kappa_1 + \mu^3) \cos(\tfrac{\pi}{3}) - 1 \right) x_2, \\
\Delta z \mid_{(x_1,x_2,z) \in \mathscr{S}} &= (\kappa_2 + \mu - 1)z,
\end{aligned} \tag{2.2.78}
$$

where $\kappa_1 = \exp(-\tfrac{\pi}{120})$, $\kappa_2 = \exp(-\tfrac{\pi}{400})$, $\mathscr{S} = s \times \mathbb{R}$, and the curve s is given by the equation $x_2 = x_1^2 + \mu x_1^3$, $x_1 > 0$. Using (2.2.76) and (2.2.77), one can define

$$
q_1(\mu) = (\kappa_1 + \mu^3) \exp\left((0.1 - \mu)\frac{5\pi}{60} \right),
$$

and

$$
q_2(\mu) = (\kappa_2 + \mu) \exp\left((-0.3 + \mu)\frac{5\pi}{60} \right).
$$

It is easily seen that $q_1(0) = \kappa_1 \exp(\tfrac{\pi}{120}) = 1$, $q_1'(0) = -\tfrac{\pi}{12} \neq 0$ and $q_2(0) = \exp(-\tfrac{11\pi}{200}) < 1$. Therefore, by Theorem 2.2.3, system (2.2.78) has a periodic solution with period $\approx \tfrac{5\pi}{60}$ if $\mid \mu \mid$ is sufficiently small.

Figure 2.7 shows the trajectory of (2.2.78) with the parameter $\mu = 0.05$ and the initial value $(x_{10}, x_{20}, z_0) = (0.02, 0, 0.05)$. Since there is an asymptotically stable center manifold, no matter which initial condition is taken, the trajectory will get closer and closer to the center manifold as time increases.

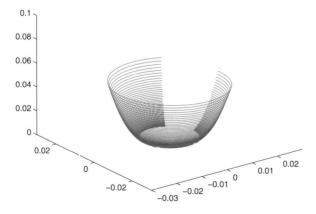

Fig. 2.7 A trajectory of (2.2.78)

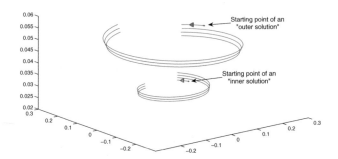

Fig. 2.8 There must exist a discontinuous limit cycle of (2.2.78)

In Fig. 2.8, the existence of a discontinuous limit cycle is illustrated. There an outer and an inner solution are shown which spiral to a trajectory lying between these two. Since the exact value of the initial point for the periodic solution is not known, we have shown two trajectories of (2.2.78).

Example 2.2.2 Consider the following dynamical system:

$$
\begin{aligned}
&x_1' = (-2 + \mu)x_1 - x_2 + \mu z^2, \\
&x_2' = x_1 + (-2 + \mu)x_2, \\
&z' = (-1 + \mu)z + \mu^2 x_1 z, \quad (x_1, x_2, z) \notin \mathscr{S}, \\
&\Delta x_1 \mid_{(x_1, x_2, z) \in \mathscr{S}} = \left((\kappa_1 - x_1^2 - x_2^2)\cos(\tfrac{\pi}{3}) - 1\right)x_1 - (\kappa_1 - x_1^2 - x_2^2)\sin(\tfrac{\pi}{3})x_2, \\
&\Delta x_2 \mid_{(x_1, x_2, z) \in \mathscr{S}} = (\kappa_1 - x_1^2 - x_2^2)\sin(\tfrac{\pi}{3})x_1 + \left((\kappa_1 - x_1^2 - x_2^2)\cos(\tfrac{\pi}{3}) - 1\right)x_2, \\
&\Delta z \mid_{(x_1, x_2, z) \in \mathscr{S}} = (\kappa_2 - 1 - z^2)z,
\end{aligned}
\tag{2.2.79}
$$

where $\kappa_1 = \exp(\frac{10\pi}{3})$, $\kappa_2 = \exp(\frac{5\pi}{6})$, $\mathscr{S} = s \times \mathbb{R}$, and s is a curve given by the equation $x_2 = x_1 + \mu^2 x_1^3$, $x_1 > 0$. Using (2.2.76) and (2.2.77), one can define

$$
q_1(\mu) = \kappa_1 \exp\left((-2 + \mu)\frac{5\pi}{3}\right),
$$

and

$$
q_2(\mu) = \kappa_2 \exp\left((-1 + \mu)\frac{5\pi}{3}\right).
$$

Now, $q_1(0) = \kappa_1 \exp(-\frac{10\pi}{3}) = 1$, $q_1'(0) = \frac{5\pi}{3} \neq 0$, $q_2(0) = \kappa_2 \exp(\frac{5\pi}{3}) = \exp(-\frac{5\pi}{6})$. Moreover, associated D-system is:

$$x_1' = -2x_1 - x_2,$$
$$x_2' = x_1 - 2x_2,$$
$$z' = -z, \quad (x_1, x_2, z) \notin \mathscr{P},$$

$$(2.2.80)$$

$$\Delta x_1 \mid_{(x_1, x_2, z) \in \mathscr{P}} = \left((\kappa_1 - x_1^2 - x_2^2) \cos(\tfrac{\pi}{3}) - 1 \right) x_1 - (\kappa_1 - x_1^2 - x_2^2) \sin(\tfrac{\pi}{3}) x_2,$$
$$\Delta x_2 \mid_{(x_1, x_2, z) \in \mathscr{P}} = (\kappa_1 - x_1^2 - x_2^2) \sin(\tfrac{\pi}{3}) x_1 + \left((\kappa_1 - x_1^2 - x_2^2) \cos(\tfrac{\pi}{3}) - 1 \right) x_2,$$
$$\Delta z \mid_{(x_1, x_2, z) \in \mathscr{P}} = (\kappa_2 - 1 - z^2) z,$$

where $\mathscr{P} = \ell \times \mathbb{R}$, ℓ is given by the equation $x_2 = x_1, x_1 > 0$, and the origin is stable focus. Indeed, using cylindrical coordinates, denote the solution of (2.2.80) starting at the angle $\phi = \frac{\pi}{4}$ by $(r(\phi, r_0, z_0), z(\phi, r_0, z_0))$.

We obtain

$$r_n = (\kappa_1 - r_{n-1}^2) r_{n-1} \exp\left(-\frac{10\pi}{3} \right),$$

and

$$z_n = (\kappa_2 - z_{n-1}^2) z_{n-1} \exp\left(-\frac{5\pi}{3} \right),$$

where $r_n = r(\frac{\pi}{4} + 2\pi n, r_0, z_0)$ and $z_n = z(\frac{\pi}{4} + 2\pi n, r_0, z_0)$. It is easily seen that the sequences r_n and z_n are monotonically decreasing for sufficiently small (r_0, z_0), and there exists a limit of (r_n, z_n). Assume that this limit is $(\xi, \eta) \neq (0, 0)$. Then, it implies that there exists a periodic solution of (2.2.80) and $\xi = (\kappa_1 - \xi^2) \xi \exp(-\frac{10\pi}{3})$ and $\eta = (\kappa_2 - \eta^2) \eta \exp(-\frac{5\pi}{3})$ which give us a contradiction. Thus, $(\xi, \eta) = (0, 0)$. Consequently, the origin is a stable focus of (2.2.80), and by Theorem 2.2.3, the system (2.2.79) has a limit cycle with period $\approx \frac{5\pi}{3}$ if $\mid \mu \mid$ is sufficiently small.

2.3 Periodic Solutions of the Van der Pol Equation

In this section, we apply the methods of B-equivalence and ψ-substitution to prove the existence of discontinuous limit cycle for the Van der Pol equation with impacts on surfaces. The result is extended through the center manifold theory for coupled oscillators. The main novelty of the result is that the surfaces, where the jumps occur, are not flat. Examples and simulations are provided to demonstrate the theoretical results as well as application opportunities.

2.3.1 Introduction and Preliminaries

Getting bifurcation in dynamics with impacts relies mainly on collisions near the impact point(s). That is why they are called corner-collision, border-collision, crossing-sliding, grazing-sliding, switching-sliding, etc., bifurcations [64, 66, 86, 112, 116, 133, 145, 181]. That is, the bifurcations are located geometrically. In our present result, we do not have the geometrical source of bifurcation. It is rather reasoned by specifically arranged interaction of continuous and discontinuous stages of the process. To be precise, we use a generalized eigenvalue to evaluate which we apply a characteristic of the impact as well as of the continuous process between moments of discontinuity. This approach when continuous and discontinuous stages are equally participated in creating a certain phenomena is common for the theory of differential equations with impulses [1, 216]. Our results are, rather, close to those, which obtained for systems where continuous flows and surfaces of discontinuity are transversal [1, 6, 39, 109, 150].

The main instruments in this section, except for the Hopf bifurcation technique, are the methods of B-equivalence and ψ-substitution developed in papers [1, 2, 6, 38, 40] for discontinuous limit cycles, and one has to emphasize that the set of all periodic solutions of the nonperturbed system is a proper subset of all solutions near the origin. By a discontinuous cycle, we mean a trajectory of a discontinuous periodic solution.

The Van der Pol equation arises in the study of circuits containing vacuum tubes and is given by

$$y'' + \varepsilon(1 - y^2)y' + y = 0 \qquad (2.3.81)$$

where ε is a real parameter. If $\varepsilon = 0$, the equation reduces to the equation of simple harmonic motion $y'' + y = 0$. The term $\varepsilon(1 - y^2)y'$ in (2.3.81) is usually regarded as the friction or resistance. If the coefficient $\varepsilon(1 - y^2)$ is positive, then we have the case of "positive resistance," and when the coefficient $\varepsilon(1 - y^2)$ is negative, then we have the case of "negative resistance." This equation, introduced by Lord Rayleigh (1896), was studied by Van der Pol (1927) [229] both theoretically and experimentally using electric circuits.

Hopf bifurcation is an attractive subject of analysis for mathematicians as well as for mechanics and engineers [6, 39, 63, 66, 84, 110, 112, 133, 148, 160, 171, 190, 228, 229]. Many papers and books have been published about mechanical and electrical systems with impacts [59, 64, 86, 116, 119, 161, 181, 232].

We consider the model with impulses on surfaces which are places in the phase space and are essentially nonlinear while it is known that the Hopf bifurcation is considered either with linear surfaces of discontinuity or with fixed moments of impulses [59, 63, 64, 66, 77, 109, 112, 116, 133, 209]. We have developed a special effective approach to analyze the problem in depth which consists of the method of reduction in equations with variable moments of impacts to systems with fixed moments of impacts [1], a class of equations on variable timescales [6, 40], and a transformation of equations on time scales to systems with impulses [38]. This is all the theoretical basis of the present results.

Specifically, we consider the following system:

$$y'' + 2\alpha y' + (\alpha^2 + \beta^2)y = F(y, y', \mu), \quad (y, y') \notin \Gamma(\mu),$$
$$\Delta y'|_{(y,y')\in\Gamma(\mu)} = cy + dy' + J(y, y', \mu), \tag{2.3.82}$$

where α, $\beta \neq 0$, c, d are real constants with $c = \alpha d$, and F and J are analytic functions in all variables. $\Gamma(\mu)$ is the set of discontinuity whose equation is given by $m_1 y + m_2 y' + \tau(y, y', \mu) = 0$, $y > 0$, for some real numbers m_1, m_2; the function $\tau(y, y', \mu)$ stands for a small perturbation; and $\Delta y'|_{(y,y')\in\Gamma(\mu)} = y'(\theta^+) - y'(\theta)$ denotes the jump operator in which θ is the time when the solution (y, y') meets the discontinuity set $\Gamma(\mu)$, that is, θ is such that $m_1 y(\theta) + m_2 y'(\theta) + \tau(y(\theta), y'(\theta), \mu) = 0$, and $y'(\theta^+)$ is the right limit of $y'(t)$ at $t = \theta$. After the impact, the phase point $(y(\theta^+), y'(\theta^+))$ will belong to the set $\Gamma'(\mu) = \{(u, v) \in \mathbb{R}^2 : u = y, v = cy + (1+d)y' + J(y, y', \mu), (y, y') \in \Gamma(\mu)\}$. Here, $y(\theta^+)$ is the right limit of $y(t)$ at $t = \theta$. One can easily see that nonlinearity is inserted into all parts of the model including the surface of discontinuity.

If we choose $\alpha = \varepsilon/2$, $\beta = \sqrt{1 - \alpha^2}$ and $F(y, y', \mu) = \varepsilon y^2 y'$ in the differential equation of the system (2.3.82), then the *Van der Pol equation* will be obtained. Therefore, (2.3.81) is a special case of (2.3.82), if the impulsive condition is not considered. Note that if $F(y, y', \mu) = \varepsilon_2 y^2 y'$ for some nonzero constant ε_2, we still have (2.3.81) after using the linear transformation $y = \sqrt{\varepsilon/\varepsilon_2}z$ of the dependent variable.

To explain our application motivations, we consider the oscillator which is subdued to the impacts modeled by the Newton's law of restitution as a concrete mechanical problem. Consider the system

$$y'' + \varepsilon_1 y' + y = \varepsilon_2 y^2 y', \quad (y, y') \notin \Gamma,$$
$$\Delta y'|_{(y,y')\in\Gamma} = dy', \tag{2.3.83}$$

where ε_1, ε_2 are constants, $d = e^{2\pi\varepsilon_1(4-\varepsilon_1^2)^{-1/2}} - 1$, Γ is the half-line $y = 0$, $y' > 0$. As it said above, the last system is a generalization of the Van der Pol equation with impacts of Newton's type. If one takes (2.3.83) with $\varepsilon_2 = 0$, then the system is

$$y'' + \varepsilon_1 y' + y = 0, \quad (y, y') \notin \Gamma,$$
$$\Delta y'|_{(y,y')\in\Gamma} = dy'. \tag{2.3.84}$$

Note that the general solution of the differential equation without impulse condition in (2.3.84) is given by

$$y(t) = e^{-\varepsilon_1 t/?} \left(C_1 \cos\left((4 - \varepsilon_1^2)^{1/2}t/2\right) + C_2 \sin\left((4 - \varepsilon_1^2)^{1/2}t/2\right)\right), \tag{2.3.85}$$

where C_1 and C_2 are arbitrary real constants. Let $(0, y_0')$ be any point on the line $\Gamma' = \Gamma$. That is, assume that $y_0' > 0$. Then, $y(0) = 0$, $y'(0) = y_0'$ in (2.3.85) gives us $C_1 = 0$, $C_2 = 2y_0'(4 - \varepsilon_1^2)^{-1/2}$. Thus, we obtain

$$y(t) = 2y_0' \left(4 - \varepsilon_1^2\right)^{-1/2} e^{-\varepsilon_1 t/2} \sin\left((4 - \varepsilon_1^2)^{1/2} t/2\right).$$

Now, the first impact action takes place at time $t = T$ where $T > 0$ and $y(T) = 0$, which means $T = 4\pi(4 - \varepsilon_1^2)^{-1/2}$. At that time, we have $y'(T) = e^{-2\pi\varepsilon_1(4-\varepsilon_1)^{-1/2}} y_0'$, and after the impact, we have $y'(T^+) = (1 + d)y(T) = y_0'$. Therefore, *all* solutions starting on Γ' are $T = 4\pi(4 - \varepsilon_1^2)^{-1/2}$ periodic. One such solution with $y_0' = 0.06$ is depicted in Fig. 2.11.

The obtained result for (2.3.84) shows that the origin is the center for the system, and it is analogous of the planar degenerated linear homogeneous system with constant coefficients in the original Hopf bifurcation theorem. This gives us a hint to apply the bifurcation technique to more general systems of type (2.3.82).

Systems of type (2.3.83) has been analyzed in many papers and books [59, 63, 66, 110, 148, 160, 216, 229] and references cited there. Here, we have mentioned just some of them. We will apply the results of current chapter to prove the existence of a stable periodic motion of the model, in the perturbed system corresponding to this model. Moreover, in Example 2.3.2, we will handle a more complicated case of two coupled oscillators where one of the oscillators is subdued to the impacts modeled by the Newton's law of restitution.

We strictly believe that results of the present section can be applied to other mechanical, electrical, as well as biological problems if one adopts the models by special transformations to the considered case. Moreover, in the upcoming researches, we plan to weaken some restrictions on the model. For example, the approach can be extended to equations where surfaces of discontinuity do not intersect at the origin.

Finally, in the present study, we extend the results to the two-oscillator model through the application of center manifold.

The analysis developed in this chapter can be applied to various problems of mechanics, electronics, and biology.

2.3.2 Theoretical Results

2.3.2.1 Reduction to Polar Coordinates

Assume that functions F, J and τ are analytic in all variables,

$$F(0, 0, \mu) = J(0, 0, \mu) = \tau(0, 0, \mu) = 0$$

for all μ, and first derivatives of F, J and τ at $(y, y', \mu) = (0, 0, 0)$ vanish. We start the theoretical investigation by writing (2.3.82) in the following form:

$$\begin{aligned}
y'' + 2\alpha(\mu)y' + (\alpha^2(\mu) + \beta^2(\mu))y &= G(y, y', \mu), \quad (y, y') \notin \Gamma(\mu), \\
\Delta y'|_{(y,y')\in\Gamma(\mu)} &= cy + dy' + J(y, y', \mu),
\end{aligned} \qquad (2.3.86)$$

where $\alpha(\mu) = \alpha - \frac{1}{2}\frac{\partial F}{\partial y'}(0, 0, \mu)$, $\beta(\mu) = \left(\alpha^2 + \beta^2 - \alpha^2(\mu) - \frac{\partial F}{\partial y}(0, 0, \mu)\right)^{1/2}$,

$G(y, y', \mu) = F(y, y', \mu) - y\frac{\partial F}{\partial y}(0, 0, \mu) - y'\frac{\partial F}{\partial y'}(0, 0, \mu)$. Note that the functions G, $\frac{\partial G}{\partial y}$ and $\frac{\partial G}{\partial y'}$ vanish at $(0, 0, \mu)$ for all μ. $\Gamma(\mu)$ can also be written as

$$\Gamma(\mu) : m_1(\mu)y + m_2(\mu)y' + \tau_2(y, y', \mu) = 0,$$

where $m_1(\mu) = m_1 + \partial\tau/\partial y(0, 0, \mu)$, $m_2(\mu) = m_1 + \partial\tau/\partial y'(0, 0, \mu)$, and $\tau_2(y, y', \mu) = \tau(y, y', \mu) - y\partial\tau/\partial y(0, 0, \mu) - y'\partial\tau/\partial y'(0, 0, \mu)$.

We write (2.3.86) as a system of first-order equations in x_1x_2-plane so that the linear part has the coefficient matrix in Jordan form. For this purpose, we let $x_1 = (\alpha(\mu)y + y')/\beta(\mu)$ and $x_2 = y$. Then, (2.3.86) is written as

$$\begin{aligned}
x_1' &= -\alpha(\mu)x_1 - \beta(\mu)x_2 + H(x_1, x_2, \mu), \\
x_2' &= \beta(\mu)x_1 - \alpha(\mu)x_2, \quad (x_1, x_2) \notin \Gamma(\mu), \\
\Delta x_1|_{(x_1, x_2)\in\Gamma(\mu)} &= Ix_1 + K(x_1, x_2, \mu),
\end{aligned} \qquad (2.3.87)$$

where $I = d$, functions H and K are analytic in all their variables and they carry all the properties of G and J, respectively. The discontinuity surface $\Gamma(\mu)$ is given by

$$\Gamma(\mu) : m_2(\mu)\beta(\mu)x_1 + (m_1(\mu) - \alpha(\mu)m_2(\mu))x_2 + \tau_3(x_1, x_2, \mu) = 0.$$

Note that system (2.3.87) is more convenient to use the polar coordinates.

We shall now introduce the polar coordinates, but first, consider the set of discontinuity points $\Gamma(\mu)$ in polar coordinates. Using the change of variables $x_1 = r\cos\phi$, $x_2 = r\sin\phi$, the curve $\Gamma(\mu)$ is represented as

$$\Gamma(\mu) : \phi = \phi_0(\mu) + \nu(r, \phi, \mu),$$

where $\phi_0(\mu) = \arctan(m_2(\mu)\beta(\mu)/(\alpha(\mu)m_2(\mu) - m_1(\mu)))$, ν is analytic in all variables, 2π-periodic in ϕ, and $\nu = \mathcal{O}(r)$. Thus, using the polar coordinates, (2.3.87) is transformed into the system

$$\begin{aligned}
r' &= -\alpha(\mu)r + R_1(r, \phi, \mu), \\
\phi' &= \beta(\mu) + R_2(r, \phi, \mu), \quad (r, \phi) \notin \Gamma(\mu), \\
\Delta r|_{(r,\phi)\in\Gamma(\mu)} &= k(\mu)r + R(r, \phi, \mu), \\
\Delta\phi|_{(r,\phi)\in\Gamma(\mu)} &= -\theta(\mu) + \Theta(r, \phi, \mu),
\end{aligned} \qquad (2.3.88)$$

where R_1, R_2, R, Θ are all 2π-periodic in ϕ, $R_1 = \mathcal{O}(r^2)$, $R_2 - \mathcal{O}(r)$, $R = \mathcal{O}(r^2)$, $\Theta = \mathcal{O}(r)$ and

$$k(\mu) = \sqrt{(I^2 + 2I)\cos^2(\phi_0(\mu)) + 1} - 1.$$

Eliminating the time variable, and considering ϕ as the independent variable, we can write (2.3.88) as

$$\frac{dr}{d\phi} = \lambda(\mu)r + \mathscr{R}(r, \phi, \mu), \quad \phi \neq \phi_0(\mu) + \xi_{2j}(r, \phi, \mu),$$
$$\Delta r|_{\phi = \phi_0(\mu) + \xi_{2j}(r, \phi, \mu)} = k(\mu)r + R(r, \phi, \mu), \quad (2.3.89)$$
$$\Delta \phi|_{\phi = \phi_0(\mu) + \xi_{2j}(r, \phi, \mu)} = -\theta(\mu) + \Theta(r, \phi, \mu),$$

where $\lambda(\mu) = -\alpha(\mu)/\beta(\mu)$, $\xi_{2j}(r, \phi, \mu) = v(r, \phi, \mu)$. In a neighborhood of the origin, it is easily seen that all solutions of (2.3.89), except for the trivial solution, rotate around the origin. Note that the impacts occur once in every two meetings of the trajectory with the discontinuity set $\Gamma(\mu)$. To indicate this notion, we use $2j$ in the subscript in (2.3.89).

During one rotation around the origin, if a solution $r(\phi)$ of (2.3.89) performs the first impact at the moment when $\phi = \eta_{2j}$, that is, $\eta_{2j} = \phi_0 + \xi_{2j}(r(\eta_{2j}), \eta_{2j}, \mu)$, and if it jumps to the point $(r(\gamma_{2j}), \gamma_{2j})$ after the impact, where $\gamma_{2j} = \eta_{2j} - \theta(\mu) + \Theta(r(\eta_{2j}), \eta_{2j}, \mu)$, then this solution is defined on the variable timescale $\cup_{j \in \mathbb{Z}}(\gamma_{2j} + 2j\pi, \eta_{2j} + 2(j+1)\pi]$. The variable timescale depends on the initial data, and the timescale is different for different solutions. Thus, (2.3.89) is considered as an impulsive differential equation on variable timescale [40].

2.3.2.2 The B-Equivalent System

In this part, we shall reduce the system in polar coordinates on variable timescale (2.3.89) to the system on the nonvariable timescale with transition condition [38], using the method of B-equivalence [1, 6, 40].

Let $r(\phi)$ be a solution of (2.3.89) with initial condition $r(\phi_0(\mu)) = r$, and assume that $\phi = \eta_{2j}$ is the first from left solution of $\phi = \phi_0(\mu) + \xi_{2j}(r, \phi, \mu)$. That is, assume that $r(\phi)$ performs the first impact at the moment when $\phi = \eta_{2j}$. Let the solution $r(\phi)$ jump to the point $(r(\gamma_{2j}), \gamma_{2j})$ after the impact. Then, we have

$$\gamma_{2j} = \eta_{2j} - \theta(\mu) + \Theta(r(\eta_{2j}), \eta_{2j}, \mu),$$
$$r(\gamma_{2j}) = (1 + k(\mu))r(\eta_{2j}) + R(r(\eta_{2j}), \eta_{2j}, \mu).$$

Throughout this section, $\widehat{[a, b]}$ denotes the oriented interval for any $a, b \in \mathbb{R}$. That is, it denotes $[a, b]$ when $a < b$, and it denotes $[b, a]$ otherwise.

Denote by $r_1(\phi)$ the solution of

$$\frac{dr}{d\phi} = \lambda(\mu)r + \mathscr{R}(r, \phi, \mu) \qquad (2.3.90)$$

with the initial condition $r_1(\gamma_{2j}) = r(\gamma_{2j})$.

For $\phi \in \widehat{[\phi_0(\mu), \eta_{2j}]}$, we have

$$r(\phi) = r + \int_{\phi_0(\mu)}^{\phi} [\lambda(\mu)r(s) + \mathscr{R}(r(s), s, \mu)]ds,$$

and on the interval $[\gamma_{2j}, \widehat{\phi_0(\mu)} - \theta(\mu)]$, we have

$$r_1(\phi) = r(\gamma_{2j}) + \int_{\gamma_{2j}}^{\phi} [\lambda(\mu)r_1(s) + \mathscr{R}(r_1(s), s, \mu)]ds.$$

Thus,

$$r_1(\phi_0(\mu) - \theta(\mu)) = r(\gamma_{2j}) + \int_{\gamma_{2j}}^{\phi_0(\mu)-\theta(\mu)} [\lambda(\mu)r_1(s) + \mathscr{R}(r_1(s), s, \mu)]ds$$

$$= (1 + k(\mu))r(\eta_{2j}) + \mathscr{R}(r(\eta_{2j}), \eta_{2j}, \mu)$$

$$+ \int_{\gamma_{2j}}^{\phi_0(\mu)-\theta(\mu)} [\lambda(\mu)r_1(s) + \mathscr{R}(r_1(s), s, \mu)]ds$$

$$= (1 + k(\mu)) \left[r + \int_{\phi_0(\mu)}^{\eta_{2j}} [\lambda(\mu)r(s) + \mathscr{R}(r(s), s, \mu)]ds \right]$$

$$+ \mathscr{R}(r(\eta_{2j}), \eta_{2j}, \mu)$$

$$+ \int_{\gamma_{2j}}^{\phi_0(\mu)-\theta(\mu)} [\lambda(\mu)r_1(s) + \mathscr{R}(r_1(s), s, \mu)]ds.$$

We now let

$$W(r, \mu) = r_1(\phi_0(\mu) - \theta(\mu)) - (1 + k(\mu))r$$

$$= (1 + k(\mu)) \int_{\phi_0(\mu)}^{\eta_{2j}} [\lambda(\mu)r(s) + \mathscr{R}(r(s), s, \mu)]ds$$

$$+ \mathscr{R}(r(\eta_{2j}), \eta_{2j}, \mu)$$

$$+ \int_{\gamma_{2j}}^{\phi_0(\mu)-\theta(\mu)} [\lambda(\mu)r_1(s) + \mathscr{R}(r_1(s), s, \mu)]ds.$$

Defining $W(u, \mu)$ and $W(v, \mu)$, using the smallness of the right side function $\mathscr{R}$ and continuous dependence on initial data, we can show that there exists a Lipschitz constant ℓ and a bounded function $m(\ell)$ such that

$$|W(u, \mu) - W(v, \mu)| \leq \ell m(\ell)|u - v| \qquad (2.3.91)$$

for all $u, v \in \mathbb{R}$. In Fig. 2.9, the construction of W is demonstrated. There, the point A is $(r, \phi_0(\mu))$, B is the point where the first impact occurs. That is, B is the point $(r(\eta_{2j}), \eta_{2j})$. The phase point jumps to C after the impact. That is, C is the point $(r(\gamma_{2j}), \gamma_{2j})$. Finally, D is the point $(r_1(\phi_0(\mu) - \theta(\mu)), \phi_0(\mu) - \theta(\mu))$.

Fig. 2.9 *B*-equivalence and
the map *W*

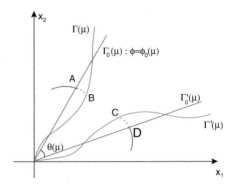

Now, we define the following system

$$
\begin{aligned}
\frac{d\rho}{d\phi} &= \lambda(\mu)\rho + \mathscr{R}(\rho, \phi, \mu), \quad \phi \neq \phi_0(\mu), \\
\Delta\rho|_{\phi=\phi_0(\mu)} &= k(\mu)\rho + W(\rho, \mu), \\
\Delta\phi|_{\phi=\phi_0(\mu)} &= -\theta(\mu).
\end{aligned}
\tag{2.3.92}
$$

It can be seen that in a neighborhood of the origin, all solutions of (2.3.92), except for the trivial solution, rotate around the origin, as in the case of (2.3.89). The variable ϕ ranges over the timescale $\cup_{n\in\mathbb{Z}}(\phi_0(\mu) - \theta(\mu) + 2n\pi, \phi_0(\mu) + 2(n+1)\pi]$. It can be seen that the timescale is a union of overlapping intervals. Indeed, (2.3.92) is an example of differential equation on a timescale with transition condition (DETCV) [38]. Nevertheless, the timescale is of a new type, since the intervals are overlapping.

Definition 2.3.1 ([6]) We say that (2.3.89) and (2.3.92) are *B*-equivalent in a neighborhood of the origin if corresponding to each solution $r(\phi)$ of (2.3.89), and there is a solution $\rho(\phi)$ of (2.3.92) such that $\rho(\phi) = r(\phi)$ for all ϕ except possibly on the intervals $[\widehat{\phi_0(\mu), \eta_{2j}}]$ and $[\widehat{\gamma_{2j}, \phi_0(\mu)} - \theta(\mu)]$, where $\eta_{2j} = \eta_{2j}(\mu)$ is the angle when $r(\phi)$ meets $\Gamma(\mu)$ and $\gamma_{2j} = \gamma_{2j}(\mu)$ is the angle where $r(\phi)$ is after the impact.

From the construction made above for *W*, one can easily see that the following lemma is valid.

Lemma 2.3.1 *Systems (2.3.89) and (2.3.92) are B-equivalent in a neighborhood of the origin.*

2.3.2.3 ψ-Substitution

The independent variable ϕ in (2.3.92) ranges over the domain $\cup_{n\in\mathbb{Z}} I_n$ where

$$
I_n = (\phi_0(\mu) - \theta(\mu) + 2n\pi, \phi_0(\mu) + 2(n+1)\pi].
$$

Note that $\cup_{n\in\mathbb{Z}} I_n$ is the union of overlapping closed intervals. That is,

$$I_{n,n+1} := I_n \cap I_{n+1} = (\phi_0(\mu) - \theta(\mu) + 2(n+1)\pi, \phi_0(\mu) + 2(n+1)\pi].$$

Therefore, we use a generalization of the so-called ψ-substitution [6, 38]. In our case, the ψ-substitution is defined to be the shifting of the intervals. More precisely, we redefine the intervals I_n as $I'_n := I_n + n\theta(\mu)$. The piece of graph of a solution is shifted accordingly. After the ψ-substitution, we obtain $I'_n \cap I'_{n+1} = \{\}$, and hence, the graph of any trajectory is a single-valued function.

Thus, using the ψ-substitution, we write (2.3.92) as

$$\begin{aligned}
\frac{d\rho}{d\varphi} &= \lambda(\mu)\rho + \tilde{\mathscr{R}}(\rho, \varphi, \mu), \quad \varphi \neq \phi_0(\mu), \\
\Delta\rho|_{\varphi=\phi_0(\mu)} &= k(\mu)\rho + \tilde{W}(\rho, \mu),
\end{aligned} \tag{2.3.93}$$

where $\varphi = \psi(\phi)$. To investigate the Hopf bifurcation in (2.3.93), we follow the classical method, and assume that the nonperturbed system has a family of periodic solutions and the origin in the perturbed system corresponding to $\mu = 0$ is asymptotically stable. Consider the case $\mu = 0$:

$$\begin{aligned}
\frac{d\rho}{d\varphi} &= \lambda\rho + \tilde{\mathscr{R}}(\rho, \varphi), \quad \varphi \neq \phi_0, \\
\Delta\rho|_{\varphi=\phi_0} &= k\rho + \tilde{W}(\rho),
\end{aligned} \tag{2.3.94}$$

where λ, k, ϕ_0, $\tilde{\mathscr{R}}(\rho, \varphi)$ and $\tilde{W}(\rho)$ are the values of $\lambda(\mu)$, $k(\mu)$, $\phi_0(\mu)$, $\tilde{\mathscr{R}}(\rho, \varphi, \mu)$ and $\tilde{W}(\rho, \mu)$ at $\mu = 0$, respectively. The nonperturbed system corresponding to (2.3.94) is

$$\begin{aligned}
\frac{d\rho}{d\varphi} &= \lambda\rho, \quad \varphi \neq \phi_0, \\
\Delta\rho|_{\varphi=\phi_0} &= k\rho.
\end{aligned} \tag{2.3.95}$$

The impacts in (2.3.95) occur on the line $\Gamma : \phi = \phi_0$, and after the impact, the trajectory is on the line $\Gamma' : \phi = \phi_0 - \theta$. The solution $r(\phi) = r(\phi, \phi_0 - \theta, r_0)$, $r_0 > 0$, of (2.3.95) with the initial condition $r(\phi_0 - \theta) = r_0$, where the point $(r_0, \phi_0 - \theta)$ is on the line Γ' (see Fig. 2.10), is given by $r(\phi) = r_0 e^{\lambda(\phi-\phi_0+\theta)}$ for $\phi_0 - \theta \leq \phi \leq \phi_0 + 2\pi$. Therefore, before the first impact, we have $r(\phi_0 + 2\pi) = r_0 e^{-\alpha T}$, where $T = (2\pi + \theta)/\beta$, and after the impact, the state position is $r(\phi_0 + 2\pi^+) = (1 + k)r(\phi_0 + 2\pi) = (1 + k)e^{-\alpha T}r_0$.

We construct the Poincaré map on the line Γ' and denote

$$q := \frac{r(\phi_0 + 2\pi^+)}{r(\phi_0 - \theta)} = (1 + k)e^{-\alpha T}, \tag{2.3.96}$$

from which we easily see that the following theorem holds.

Theorem 2.3.1 *If*

(a) *$q = 1$, then all solutions of (2.3.95) with the initial conditions on Γ' are T-periodic;*
(b) *$q < 1$, then all solutions of (2.3.95) with the initial conditions on Γ' spiral in toward the origin;*
(c) *$q > 1$, then all solutions of (2.3.95) with the initial conditions on Γ' move away from the origin.*

Remark 2.3.1 In this study, for given α, β, and θ, we fix the number I as $I = (e^{\alpha T} - \cos\theta)/(\cos\theta - e^{-\alpha T}) - 1$ so that $q = 1$ and hence part (a) of the Theorem 2.3.1 holds (see Fig. 2.10). Since $q \neq 1$ is the noncritical case, when the phase portrait is persistent under perturbations, our present interest is only with the case (a) of the Theorem 2.3.1.

Remark 2.3.2 If q, which is defined in (2.3.96), is less than 1, then any solution of (2.3.94) with the initial condition on Γ' spirals in toward the origin, and if $q > 1$, then the solutions move away from the origin. On the other hand, when $q = 1$ we have the critical case. That is, a solution of (2.3.94) with the initial condition on Γ' may be periodic, and it may spiral in toward the origin or it may move away from the origin. In this study, as mentioned before, the case $q \neq 1$ is not of our interest and the critical case is investigated below.

2.3.2.4 Hopf Bifurcation

In this section, we consider (2.3.93) again:

$$\frac{d\rho}{d\varphi} = \lambda(\mu)\rho + \tilde{\mathscr{R}}(\rho, \varphi, \mu), \quad \varphi \neq \phi_0(\mu),$$
$$\Delta\rho|_{\varphi=\phi_0(\mu)} = k(\mu)\rho + \tilde{W}(\rho, \mu). \tag{2.3.97}$$

Fig. 2.10 All solutions of (2.3.95) with the initial condition on Γ' are periodic

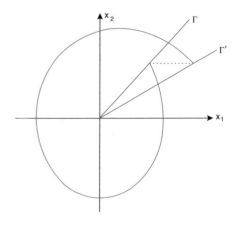

We consider the corresponding linearized system around the trivial solution:

$$\frac{d\rho}{d\varphi} = \lambda(\mu)\rho, \quad \varphi \neq \phi_0(\mu),$$
$$\Delta\rho|_{\varphi=\phi_0(\mu)} = k(\mu)\rho. \tag{2.3.98}$$

We construct the Poincaré map on the line $\Gamma_0'(\mu)$ and denote

$$q(\mu) = (1 + k(\mu))e^{-\alpha(\mu)T(\mu)}, \tag{2.3.99}$$

in the same way as we defined (2.3.96).

Theorem 2.3.2 *Assume that $q(0) = 1$, $q'(0) \neq 0$. Then for sufficiently small r_0, there exists a function $\mu = \delta(r_0)$ with $\delta(0) = 0$ such that the solution $r(\phi, r_0, \delta(r_0))$ of (2.3.89) is periodic with a period $T = (2\pi + \theta)/\beta + o(|\mu|)$. Furthermore, if solutions of (2.3.94) spiral in toward the origin, then this periodic solution is a discontinuous limit cycle.*

Proof Let $\rho(\varphi, r_0, \mu)$ be a solution of (2.3.97). Because of the analyticity of solutions [1], we have

$$\rho(2\pi, r_0, \mu) = \sum_{i=1}^{\infty} a_i(\mu)r_0^i$$

where $a_i(\mu) = \sum_{j=0}^{\infty} a_{ij}\mu^j$, $a_{10} = q(0) = 1$, $a_{11} = q'(0) \neq 0$. Define

$$\mathcal{V}(r_0, \mu) = \rho(2\pi, r_0, \mu) - r_0$$
$$= q'(0)\mu r_0 + \sum_{i=2}^{\infty} a_{i0}r_0^i + r_0\mu^2 M_1(r_0, \mu) + r_0^2\mu M_2(r_0, \mu)$$

where M_1 and M_2 are analytic functions of r_0, μ in a small neighborhood of the $(0, 0)$. When the bifurcation equation, $\mathcal{V}(r_0, \mu) = 0$, is simplified by r_0, one can write

$$\mathcal{H}(r_0, \mu) = 0 \tag{2.3.100}$$

where

$$\mathcal{H}(r_0, \mu) = q'(0)\mu + \sum_{i=2}^{\infty} a_{i0}r_0^{i-1} + \mu^2 M_1(r_0, \mu) + r_0\mu M_2(r_0, \mu).$$

By the implicit function theorem, since

$$\mathcal{H}(0, 0) = 0, \quad \frac{\partial\mathcal{H}(r_0, \mu)}{\partial\mu} = q'(0) \neq 0,$$

for sufficiently small r_0, there exists a function $\mu = \delta(r_0)$ with $\delta(0) = 0$ such that $r(\phi, r_0, \delta(r_0))$ is a periodic solution. If we assume that $a_{i0} = 0$ for $i = 2, \ldots, \ell - 1$ and $a_{\ell 0} \neq 0$, then from (2.3.100) one can obtain that

$$\delta(r_0) = -\frac{a_{\ell 0}}{q'(0)} r_0^{\ell-1} + \sum_{i=\ell}^{\infty} \delta_i r_0^i. \tag{2.3.101}$$

Analyzing the last expression, one can conclude that the bifurcation of periodic solution exists if a stable focus for $\mu = 0$ is unstable for $\mu \neq 0$ and vice versa. Let $\rho(\varphi) = \rho(\varphi, \bar{r}_0, \bar{\mu})$ be a periodic solution of (2.3.97). It is known that the trajectory is limit cycle if

$$\frac{\partial \mathcal{V}(\bar{r}_0, \bar{\mu})}{\partial r_0} < 0. \tag{2.3.102}$$

Now,

$$\frac{\partial \mathcal{V}(r_0, \mu)}{\partial r_0} = q'(0)\mu + \sum_{i=2}^{\infty} i a_{i0} r_0^{i-1} + \mu^2 N_1(r_0, \mu) + r_0 \mu N_2(r_0, \mu).$$

If $a_{\ell 0}$ is the first nonzero element among a_{i0} and $a_{\ell 0} < 0$, then using (2.3.101) one can obtain

$$\frac{\partial \mathcal{V}(\bar{r}_0, \bar{\mu})}{\partial r_0} = (\ell - 1) a_{\ell 0} \bar{r}_0^{\ell-1} + Q(\bar{r}_0),$$

where Q starts with a member whose order is not less than ℓ. Hence, (2.3.102) is valid. From the ψ-substitution and B-equivalence of (2.3.89) and (2.3.92), one can conclude that the theorem is proved.

Since the change of variables $x_1 = r \cos \phi$, $x_2 = r \sin \phi$, and $y = x_2$, $y' = \beta x_1 - \alpha x_2$ are one-to-one for $\beta \neq 0$, we see that the following theorem is valid.

Theorem 2.3.3 *Assume that $q(0) = 1$, $q'(0) \neq 0$. Then for sufficiently small initial condition $y_0 := (y(0), y'(0))$ there exists a function $\mu = \delta(y_0)$ with $\delta(0) = 0$ such that the solution $y(t, 0, y_0, \mu)$ of (2.3.86) is periodic with a period $T = (2\pi + \theta)/\beta + o(|\mu|)$. Furthermore, if solutions of (2.3.86) with the initial point on $\Gamma_0'(0)$ spiral in toward the origin, then this periodic solution is a discontinuous limit cycle.*

Example 2.3.1 In the following example, we shall consider the model which is studied in many papers and books [59, 63, 66, 110, 148, 160, 216, 229]. Here, we insert the impulse condition and consider

$$\begin{aligned} y'' + \varepsilon_1 y' + y &= \varepsilon_2 y^2 y', \quad (y, y') \notin \Gamma, \\ \Delta y'|_{(y,y')\in\Gamma} &= dy', \end{aligned} \tag{2.3.103}$$

where ε_1 and ε_2 are some nonzero real numbers and Γ is the discontinuity set which is defined, in yy'-plane, by $y = 0$, $y' > 0$. The nonperturbed system is written as

$$y'' + 2\alpha y' + (\alpha^2 + \beta^2)y = 0, \quad (y, y') \notin \Gamma,$$
$$\Delta y'|_{(y,y')\in\Gamma} = dy'. \tag{2.3.104}$$

where $\alpha = \varepsilon_1/2$, $\beta = \sqrt{1 - \alpha^2}$, $d = e^{2\pi\alpha/\beta} - 1$. Note that the general solution of the differential equation without impulse condition in (2.3.104) is given by

$$y(t) = e^{-\alpha t}(C_1 \cos(\beta t) + C_2 \sin(\beta t)), \tag{2.3.105}$$

where C_1 and C_2 are arbitrary real constants. Let $(0, y_0')$ be any point on the line $\Gamma' = \Gamma$. That is, assume that $y_0' > 0$. Then, $y(0) = 0$, $y'(0) = y_0'$ in (2.3.105) gives us $C_1 = 0$, $C_2 = y_0'/\beta$. Thus, we obtain

$$y(t) = y_0' e^{-\alpha t} \sin(\beta t)/\beta.$$

Now, the first impact action takes place at time $t = T$ where $T > 0$ and $y(T) = 0$, which means $T = 2\pi/\beta$. At that time, we have $y'(T) = e^{-2\pi\alpha/\beta} y_0'$, and after the impact, we have $y'(T^+) = (1 + d)y(T) = y_0'$. Therefore, all solutions starting on Γ' are $T = 2\pi/\beta$ periodic. One such solution with $y_0' = 0.06$ is depicted in Fig. 2.11.

The fact that for (2.3.104), the origin is a center makes it suitable for application of our theoretical results. For this reason, let us perturb (2.3.103) and bring it to the notation of system (2.3.82). That is, let us consider the model

$$y'' + 2\alpha y' + (\alpha^2 + \beta^2)y = F(y, y', \mu), \quad (y, y') \notin \Gamma(\mu),$$
$$\Delta y'|_{(y,y')\in\Gamma(\mu)} = cy + dy' + J(y, y', \mu), \tag{2.3.106}$$

Fig. 2.11 A solution of the nonperturbed system (2.3.104) with the initial condition $y(0) = 0$, $y'(0) = 0.06$

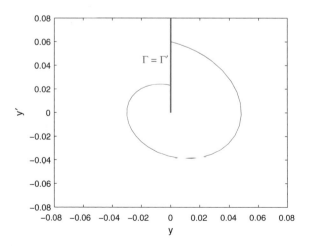

where $\alpha = 0.15$, $\beta = \sqrt{1-\alpha^2}$, $F(y, y', \mu) = 0.02\alpha y^2 y' - \mu y(2 + y + y')$, $\Gamma(\mu)$ is the curve $\Gamma(\mu) = \{(y, y') \in \mathbb{R}^2 : y + 30\mu(y')^2 = 0, y' > 0\}$, $c = \alpha d$, $d = e^{2\pi\alpha/\beta} - 1$, and $J(y, y', \mu) = -(2 + \mu)(y')^2$. Note that (2.3.106) is a special case of (2.3.82) with $m_1 = 1$, $m_2 = 0$, $\tau(y, y', \mu) = 30\mu(y')^2$. The term cy in (2.3.106) has to be considered now as a small perturbation because of smallness of y as the first coordinate of points $\Gamma(\mu)$.

To prove the existence of a periodic solution for (2.3.106), we find that the generalized eigenvalue is

$$q(\mu) = \exp\left(2\pi\alpha\left(\frac{1}{\beta} - \frac{1}{\sqrt{\beta^2 + 2\mu}}\right)\right).$$

Then, one can easily find that $q(0) = 1$ and $q'(0) \neq 0$. Therefore, by Theorem 2.3.3, for sufficiently small μ, there exists a periodic solution with period $\approx 40\pi/\sqrt{391}$.

When $\mu = 0$ in (2.3.106), the origin is a stable focus. This can be seen in simulations and the solution of (2.3.106) corresponding to $\mu = 0$ with the initial condition $y(0) = 0$, $y'(0) = 0.06$ is drawn in Fig. 2.12.

To see an application of Theorem 2.3.3, we take $\mu = 0.08$. By Theorem 2.3.3, we know that there exists a periodic solution corresponding to an initial value $(y(0), y'(0))$ in a sufficiently small neighborhood of the origin. Two solutions of (2.3.106) are drawn in Fig. 2.13. An "inner" solution with the initial condition $y(0) = 0$, $y'(0) = 0.06$ is drawn in red curve and an "outer" solution with the initial condition $y(0) = 0$, $y'(0) = 0.12$ is drawn in blue curve. These two solutions approach to the limit cycle from inside and outside, respectively. Moreover, by Theorem 2.3.3, the period of the discontinuous limit cycle is approximately $40\pi/\sqrt{391}$.

Note that the differential equation of (2.3.106) for $\mu = 0$ becomes (2.3.81) with $\varepsilon = 0.3$. This example shows the application importance of our results.

Fig. 2.12 A solution of the perturbed system with the initial condition $y(0) = 0$, $y'(0) = 0.06$

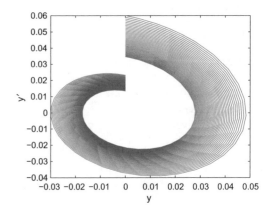

Fig. 2.13 An "inner" and an "outer" solution of (2.3.106). The inner solution is shown in *red* and it corresponds to the initial condition $y(0) = 0$, $y'(0) = 0.06$. The outer solution is shown in *blue* and it corresponds to the initial condition $y(0) = 0$, $y'(0) = 0.12$

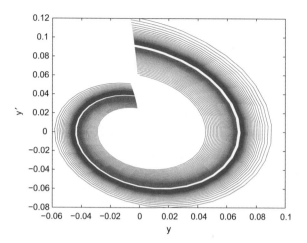

System (2.3.106) for $\mu = 0$ is one of the widely investigated models of the mechanisms with impacts determined by the Newton's law of restitution. These models have been studied in many books [66, 216] and papers cited there. One should mention that surfaces of discontinuity in these results are flat. However, it is the first time, we consider the surface of discontinuity as perturbed nonlinearly.

2.3.3 Center Manifold

In this section, we begin our development of the techniques and extend our results to coupled oscillators. We show the existence of a center manifold. Consider

$$y'' + 2\alpha y' + (\alpha^2 + \beta^2)y = F_1(y, y', z, z', \mu),$$
$$z'' + 2\gamma z' + (\gamma^2 + \sigma^2)z = F_2(y, y', z, z', \mu), \quad (y, y') \notin \Gamma(\mu), \quad (2.3.107)$$
$$\Delta y'|_{(y,y')\in\Gamma(\mu)} = cy + dy' + J(y, y', \mu),$$

where y, y' are the components of one oscillator, say (A), and z, z' are the components of another oscillator, say (B).

By using the means of the center manifold theorem [194] and its role for the Hopf bifurcation in multidimensional systems, one can predict that the system of oscillators (A) and (B) admits a limit cycle. Indeed, looking at the results of our simulations given in Figs. 2.14, 2.15, and 2.16, one can see that for the oscillator (A) we have a discontinuous limit cycle, as in Example 2.3.1, and for the oscillator (B) we have a continuous limit cycle. That is, for the whole system, the periodic trajectory (y, z) is a discontinuous limit cycle.

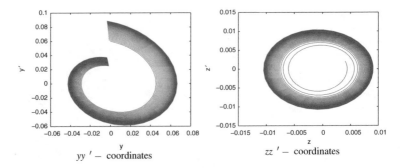

Fig. 2.14 An "inner" solution of (2.3.117) for $\mu = 0.08$ corresponding to the initial condition $y(0) = 0$, $y'(0) = 0.06$, $z(0) = 0.004$, $z'(0) = 0.002$

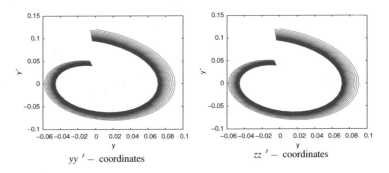

Fig. 2.15 An "outer" solution of (2.3.117) for $\mu = 0.08$ corresponding to the initial condition $y(0) = 0$, $y'(0) = 0.12$, $z(0) = 0.04$, $z'(0) = 0.02$

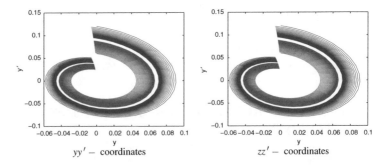

Fig. 2.16 The "inner" and "outer" solutions of (2.3.117) given in Figs. 2.14 and 2.15 are shown in the same picture.

As we did to obtain (2.3.86), we write (2.3.107) in the form

$$
\begin{aligned}
&y'' + 2\alpha(\mu)y' + (\alpha^2(\mu) + \beta^2(\mu))y = G_1(y, y', z, z', \mu),\\
&z'' + 2\gamma(\mu)z' + (\gamma^2(\mu) + \sigma^2(\mu))z = G_2(y, y', z, z', \mu), \quad (y, y') \notin \Gamma(\mu), \quad (2.3.108)\\
&\Delta y'|_{(y,y')\in\Gamma(\mu)} = cy + dy' + J(y, y', \mu).
\end{aligned}
$$

Let $y_1 = (\alpha(\mu)y + y')/\beta(\mu)$, $y_2 = y$, $z_1 = (\gamma(\mu)z + z')/\sigma(\mu)$, $z_2 = z$. Then, (2.3.108) becomes

$$
\begin{aligned}
y_1' &= -\alpha(\mu)y_1 - \beta(\mu)y_2 + H_1(y_1, y_2, z_1, z_2, \mu),\\
y_2' &= \beta(\mu)y_1 - \alpha(\mu)y_2,\\
z_1' &= -\gamma(\mu)z_1 - \sigma(\mu)z_2 + H_2(y_1, y_2, z_1, z_2, \mu), \quad\quad (2.3.109)\\
z_2' &= \sigma(\mu)z_1 - \gamma(\mu)z_2, \quad (y_1, y_2) \notin \Gamma(\mu),\\
\Delta y_1|_{(y_1,y_2)\in\Gamma(\mu)} &= Iy_1 + K(y_1, y_2, \mu).
\end{aligned}
$$

We now use the cylindrical coordinates. That is, we use the polar coordinates for the oscillator (A). Then, we eliminate the variable t, and obtain

$$
\begin{aligned}
\frac{dr}{d\phi} &= \lambda(\mu)r + \mathscr{R}(r, \phi, z_1, z_2, \mu),\\
\frac{dz_1}{d\phi} &= -\tilde\gamma(\mu)z_1 - \tilde\sigma(\mu)z_2 + \mathscr{R}_2(r, \phi, z_1, z_2, \mu),\\
\frac{dz_2}{d\phi} &= \tilde\sigma(\mu)z_1 - \tilde\gamma(\mu)z_2, \quad \phi \neq \phi_0(\mu) + \xi_{2j}(r, \phi, \mu), \quad\quad (2.3.110)\\
\Delta r|_{\phi=\phi_0(\mu)+\xi_{2j}(r,\phi,\mu)} &= k(\mu)r + R(r, \phi, \mu),\\
\Delta \phi|_{\phi=\phi_0(\mu)+\xi_{2j}(r,\phi,\mu)} &= -\theta(\mu)r + \Theta(r, \phi, \mu),
\end{aligned}
$$

where $\tilde\gamma(\mu) = \gamma(\mu)/\beta(\mu)$, $\tilde\sigma(\mu) = \sigma(\mu)/\beta(\mu)$, and all other elements except for $\mathscr{R}_2$ are the same as in (2.3.89). Using B-equivalence and ψ-substitution, we get the following system:

$$
\begin{aligned}
\frac{d\rho}{d\varphi} &= \lambda(\mu)\rho + \tilde{\mathscr{R}}(\rho, \varphi, z_1, z_2, \mu),\\
\frac{dz_1}{d\varphi} &= -\tilde\gamma(\mu)z_1 - \tilde\sigma(\mu)z_2 + \tilde{\mathscr{R}}_2(\rho, \phi, z_1, z_2, \mu),\\
\frac{dz_2}{d\varphi} &= \tilde\sigma(\mu)z_1 - \tilde\gamma(\mu)z_2, \quad \varphi \neq \phi_0(\mu), \quad\quad (2.3.111)\\
\Delta \rho|_{\varphi=\phi_0(\mu)} &= k(\mu)\rho + \tilde{W}(\rho, \mu).
\end{aligned}
$$

The corresponding linearized system around the origin is

$$\frac{d\rho}{d\varphi} = \lambda(\mu)\rho,$$
$$\frac{dz_1}{d\varphi} = -\tilde{\gamma}(\mu)z_1 - \tilde{\sigma}(\mu)z_2,$$
$$\frac{dz_2}{d\varphi} = \tilde{\sigma}(\mu)z_1 - \tilde{\gamma}(\mu)z_2, \quad \varphi \neq \phi_0(\mu), \tag{2.3.112}$$
$$\Delta\rho|_{\varphi=\phi_0(\mu)} = k(\mu)\rho.$$

Like we obtained (2.3.99), we define

$$q_1(\mu) = (1 + k(\mu))e^{-\alpha(\mu)T(\mu)},$$
$$q_2(\mu) = e^{-\tilde{\gamma}(\mu)T(\mu)}. \tag{2.3.113}$$

Note that $q_1(0) = 1$, and hence, we have the critical case for the first oscillator and $q_2(0) < 1$ if and only if $\tilde{\gamma}(0) > 0$. For our system, we assume that $q_1(0) = 1$ and $q_2(0) < 1$.

By using the formulas for the integral manifolds developed in [194], one can see that system (2.3.111) has a center manifold $S_0(\mu) := \{(\rho, \varphi, u) : u = \Phi_0(\varphi, \rho, \mu)\}$ and a stable manifold $S_-(\mu) := \{(\rho, \varphi, u) : \rho = \Phi_-(\varphi, u, \mu)\}$ where

$$\Phi_0(\varphi, \rho, \mu) = \int_{-\infty}^{\varphi} \pi_0(\varphi, s, \mu)\tilde{\mathscr{R}}_2(\rho(s, \varphi, \rho, \mu), s, u(s, \varphi, \rho, \mu), \mu)ds \tag{2.3.114}$$

and

$$\Phi_-(\varphi, u, \mu) = -\int_{\varphi}^{\infty} \pi_-(\varphi, s, \mu)\tilde{\mathscr{R}}_1(\rho(s, \varphi, u, \mu), s, u(s, \varphi, u, \mu), \mu)ds$$
$$+ \sum_{\varphi_i < \varphi} \pi_-(\varphi, \varphi_i^+, \mu)\tilde{W}(\rho(s, \varphi, u, \mu), \mu) \tag{2.3.115}$$

in which $\varphi_i = \varphi + 2\pi i$, $u = (z_1, z_2)$,

$$\pi_0(\varphi, s, \mu) = e^{-\tilde{\gamma}(\mu)(\varphi-s)}\begin{bmatrix} \cos(\tilde{\sigma}(\mu) - s) & -\sin(\tilde{\sigma}(\mu) - s) \\ \sin(\tilde{\sigma}(\mu) - s) & \cos(\tilde{\sigma}(\mu) - s) \end{bmatrix}$$

and

$$\pi_-(\varphi, s, \mu) = e^{\lambda(\mu)(\varphi-s)} \prod_{s \leq \varphi_j(\mu) < \varphi} (1 + k(\mu)).$$

In (2.3.114), the pair $(\rho(s, \phi, \rho, \mu), u(s, \phi, \rho, \mu))$ denotes a solution of (2.3.111) satisfying $\rho(s, \phi, \rho, \mu) = \rho$. Similarly, in (2.3.115), the pair $(\rho(s, \phi, u, \mu), u(s, \phi, u, \mu))$ denotes a solution of (2.3.111) satisfying $u(s, \phi, u, \mu) = u$.

On the local center manifold, $S_0(\mu)$, the first coordinate of the solutions of (2.3.111) satisfies the following system

$$
\frac{d\rho}{d\varphi} = \lambda(\mu)\rho + \tilde{\mathscr{R}}(\rho, \varphi, \Phi_0(\varphi, \rho, \mu), \mu), \quad \varphi \neq \phi_0(\mu),
$$
$$
\Delta\rho|_{\varphi=\phi_0(\mu)} = k(\mu)\rho + \tilde{W}(\rho, \mu). \tag{2.3.116}
$$

By Theorem 2.3.3, we know that, for sufficiently small μ, system (2.3.116) has a periodic solution with period $T = (2\pi + \theta)/\beta + o(\mu)$. That is, on the local center manifold $S_0(\mu)$, the ρ-coordinate of a solution of (2.3.111) is T-periodic. Because of the T-periodic properties of the right side functions, one can show that z_1 and z_2 components of a solution of (2.3.111) are also T-periodic when the ρ-coordinate is. Thus, we have the following theorem.

Theorem 2.3.4 *Assume that $q_1(0) = 1$, $q_1'(0) \neq 0$, $q_2(0) < 1$. Then for sufficiently small initial condition $(y_0, z_0) := (y(0), y'(0), z(0), z'(0))$ there exists a function $\mu = \delta(y_0, z_0)$ such that the solution $(y(t, 0, y_0, \mu), z(t, 0, z_0, \mu)$ of (2.3.107) is periodic with a period $T = (2\pi + \theta)/\beta + o(|\mu|)$. Furthermore, if solutions of (2.3.107) with the y_0-component of the initial point on $\Gamma'(0)$ spiral in toward the origin, then this periodic solution is a discontinuous limit cycle.*

Example 2.3.2 Let us develop the model studied in Example 2.3.1 further to two coupled oscillators where one of the oscillators is subdued to the impacts. For this reason, consider

$$
y'' + 2\alpha y' + (\alpha^2 + \beta^2)y = F_1(y, y', z, z', \mu),
$$
$$
z'' + 2\gamma z' + (\gamma^2 + \sigma^2)z = F_2(y, y', z, z', \mu), \quad (y, y') \notin \Gamma(\mu), \tag{2.3.117}
$$
$$
\Delta y'|_{(y,y')\in\Gamma(\mu)} = dy' + J(y, y', \mu),
$$

where $\gamma = 0.2$, $\sigma = 1$,

$$
F_1(y, y', z, z', \mu) = 0.02\alpha y^2 y' - \mu y(2 + y + z + (z')^2),
$$
$$
F_2(y, y', z, z', \mu) = \gamma yz' - \mu y(1 + z^2 + (z')^2),
$$

and all other elements are the same as in Example 2.3.1.

As evaluated before, we have $q_1(0) = 1$, $q_1'(0) \neq 0$, and $\gamma > 0$ implies that $q_2(0) < 1$. Thus, by Theorem 2.3.4, for sufficiently small μ, there exists a periodic solution with period $\approx 40\pi/\sqrt{391}$. Let $\mu = 0.08$. An "inner" solution with the initial condition $y(0) = 0$, $y'(0) = 0.06$, $z(0) = 0.004$, $z'(0) = 0.002$ is drawn in Fig. 2.14 and an "outer" solution with the initial condition $y(0) = 0$, $y'(0) = 0.12$, $z(0) = 0.04$, $z'(0) = 0.02$ is drawn in Fig. 2.15. These two solutions approach to the limit cycle from inside and outside, respectively.

2.4 Notes

The present chapter contains mainly results of papers [6, 39, 42] and is based on the perturbation theory, which was founded by H. Poincaré and A.M. Lyapunov [169, 197], and the bifurcation methods [60, 65, 121, 129, 132, 171, 173, 207, 233]. The main result is the bifurcation of a periodic solution from the equilibrium of the discontinuous dynamical system. After the initial impetus of H. Poincaré [195], A. Andronov [60], and E. Hopf [129], this method of research of periodic motions has been used very successfully for various differential equations by many authors (see [109, 121, 132, 173] and references cited there). There have been two principal obstacles of expansion of this method for discontinuous dynamical systems. While the absence of developed differentiability of solutions has been the first one, the choice of a nonperturbed system convenient to study has been the second. The present investigation utilizes extensively the differentiability and analyticity of discontinuous solutions discussed in Chap. 6 of [1]. The nonperturbed equation is specifically defined. The results of the present chapter can be extended by the dimension enlarging [39] and application to differential equations with discontinuous right side [17]. They are applied to control the population dynamics [19], and can be effectively employed in mechanics, electronics, biology, and medicine [60, 70, 121, 173, 178, 185].

In the second section, we have studied the existence of a center manifold and the Hopf bifurcation for a certain three-dimensional discontinuous dynamical system. The bifurcation of discontinuous cycle is observed by means of the B-equivalence method and its consequences. These results will be extended to arbitrary dimension for a more general type of equations.

Many evolutionary processes are subject to the short-term perturbation whose duration is negligible when compared to that of the whole process. This perturbation results in a change in the state of the process. This change can be at fixed moments or when the state process meets a certain set of discontinuity. These systems model a variety of problems of mechanics, electronics, physics, chemistry, medicine, etc., [59, 63, 64, 66, 77, 84, 109, 112, 116, 130, 145, 150, 179, 209, 228, 232].

In most of the references cited here, the impulse action or the change in the phase space takes place on a flat surface. The theory in which nonlinear surfaces of discontinuity are present has not been investigated fully because of the lack of theoretical results. The problem of discontinuous models where the surfaces of discontinuities are not flat is very actual because of natural possibilities of perturbations. It is natural that one should involve perturbation not only into the differential equation or in the impulse function, but also into the equations for the surfaces of discontinuity.

The discontinuities of the equation which determines the moments of jumps are investigated in many papers and books [216]. In most general form, the results are formulated and expressed in [1], and the present results widely use this information.

In last section, the method of B-equivalence and ψ-substitution [1, 2, 6, 38, 40] is used effectively to observe the Hopf bifurcation of periodic solution. We proved the existence of discontinuous limit cycle for the Van der Pol equation performing impacts on surfaces. We extended the results to two coupled oscillators through the application of the center manifold theory [194]. These theoretical results could be extended to arbitrary dimension and apply them to well-known discontinuous mechanical models.

One should mention that we consider the Hopf bifurcation without reduction in the problem of Hopf bifurcation of the maps as it is usually done in the literature. So, the system which admits the origin as a center in our case is nonlinear one (2.3.95), but its elements are linear. This approach to the investigation is respectively new and promisive.

Based on the present results, one can investigate multioscillatory system where not only one of the oscillators is discontinuous but several of them are discontinuous [63, 114, 118, 190].

Chapter 3
Hopf Bifurcation in Filippov Systems

3.1 Nonsmooth Planar Limit Cycle from a Vertex

We investigate nonsmooth planar systems of differential equations with discontinuous right-hand side. Discontinuity sets intersect at a vertex and are of the quasilinear nature. By means of the B-equivalence method, which was introduced in [46, 54, 55] (see also [1, 6]), these systems are reduced to impulsive differential equations. Sufficient conditions are established for the existence of foci and centers both in the noncritical and critical cases. Hopf bifurcation is considered from a vertex, which unites several curves, in the critical case. An appropriate example is provided to illustrate the results.

3.1.1 Introduction

The theory of differential equations with discontinuous right-hand side has been substantially developed through numerous applications. There are many problems from mechanics, engineering sciences [59, 146, 147, 173], control theory [115], and economics [127] that are modeled by dynamical systems with discontinuous vector fields. Besides, the books [59, 65, 173], which concern mechanical systems with dry friction, periodic solutions of discontinuous systems, and discontinuous oscillations, form an important basis for the development of such discontinuous systems. Owing to the problems of applied nature, qualitative theory of classical ordinary differential equations including the notions of existence, uniqueness, continuous dependence, stability, and bifurcation has been carefully adapted for equations with discontinuous right-hand side. The main trends of the theory can be found in [115].

Bifurcations in smooth systems are well understood [72, 89, 121, 171], but little is known in discontinuous systems. Stimulated by nonsmooth phenomena in the

© Springer Nature Singapore Pte Ltd. and Higher Education Press 2017
M. Akhmet and A. Kashkynbayev, *Bifurcation in Autonomous
and Nonautonomous Differential Equations with Discontinuities*,
Nonlinear Physical Science, DOI 10.1007/978-981-10-3180-9_3

real world, subject of Hopf bifurcation in discontinuous systems has received great attention in recent years [92, 146, 147, 150, 166, 173, 237].

Most of the papers in the literature assume that discontinuity sets of nonsmooth systems consist of a single surface, especially a straight line [92, 146, 150, 237]. However, due to exterior effects, discontinuities may appear on curves or surfaces of nonlinear feature. Hence, it is reasonable to perturb the sets of discontinuities. Kunze [146] and Küpper et al. [150, 237] address bifurcation of periodic solutions for planar Filippov systems with discontinuities on a single straight line. In [237], generalized Hopf bifurcation for the following piecewise smooth planar system

$$\begin{pmatrix} x' \\ y' \end{pmatrix} = \begin{cases} f^+(x, y, \lambda), \ x > 0, \\ f^-(x, y, \lambda), \ x < 0, \end{cases}$$

where $f^\pm(x, y, \lambda) = A^\pm(\lambda)(x, y)^T + g^\pm(x, y, \lambda)$, λ a real parameter, has been investigated using differential inclusions. Eigenvalues of the matrix $A^\pm(\lambda)$ were assumed to be complex conjugate, i.e., $\alpha^\pm(\lambda) \pm i\omega^\pm(\lambda)$. This system has been stimulated by a brake system of the form

$$mu'' + d_1u' + c_1u = \sigma^+(u, u', \lambda), \qquad\qquad\quad \text{if } u > 0,$$
$$mu'' + (d_1 + d_2)u' + (c_1 + c_2)u = \sigma^-(u, u', \lambda), \ \text{if } u < 0,$$

where a mass m rests on a smooth surface and is connected to the walls by springs (c_1 and c_2) and dampers (d_1 and d_2). $\sigma^\pm$ denotes the external force, and the parameter λ controls its magnitude (see [237] for details).

In papers [6, 236], possibly for the first time, a special structure of the domain has been developed for planar differential equations with discontinuities. To say more clearly, [6] treats bifurcation of periodic solutions for planar discontinuous dynamical systems where discontinuities in the state variable appear on countably many curves intersecting at the origin, and [236] studies generalized Hopf bifurcation for piecewise smooth planar systems with discontinuities on several straight lines emanating from the origin. We suppose that domains of this type can be useful in mechanical and electrical models with discontinuities under proper transformations.

Our present work is an attempt to generalize the problem of Hopf bifurcation for a planar nonsmooth system by considering discontinuities on finitely many nonlinear curves emanating from a vertex. We consider the domain in a neighborhood of a vertex which unites several curves. That is, the phase space is divided into subdomains and the system is described by a different set of differential equations in each domain. We can say that the system considered in this chapter is more general than the one in [236], where discontinuities occur at straight lines. We aim to give some theoretical backgrounds rather than applications, which will be very useful in many problems in the future. Using B-equivalence [1, 6, 54, 55] of the issue systems to impulsive differential equations, we obtain corresponding qualitative properties. It is the inherent advantage of the B-equivalence method that we can study equations with nonlinear discontinuity sets.

This section is organized in the following way. In Sect. 3.1.2, we introduce the non-perturbed system and study existence of foci and centers for that system. Section 3.1.3 presents the perturbed system and the notion of B-equivalent impulsive systems. The problem of distinguishing between the center and the focus is solved in Sect. 3.1.4. We investigate bifurcation of periodic solutions in the next section. Afterward, an appropriate example is worked out to illustrate our results.

3.1.2 The Nonperturbed System

Let $\mathbb{N} = \{1, 2, \ldots\}$ and $\mathbb{R}$ be the sets of natural and real numbers, respectively. Let $\mathbb{R}^2$ be the two-dimensional real space and $\langle x, y \rangle$ denote the scalar product for all vectors $x, y \in \mathbb{R}^2$. The norm of a vector $x \in \mathbb{R}^2$ is given by $\|x\| = \langle x, x \rangle^{\frac{1}{2}}$.

For the sake of brevity in the sequel, every angle for a point is considered with respect to the positive half-line of the first coordinate axis.

In the rest of the present section, following assumptions will be needed.

(A1)   Let $\{l_i\}_{i=1}^p$, $p \geq 2$, $p \in \mathbb{N}$, be a set of half-lines starting at the origin and given by the equations $\Phi_i(x) = 0$, $\Phi_i(x) = \langle a^i, x \rangle$, $i = 1, 2, \ldots, p$, where $a^i = (a_1{}^i, a_2{}^i) \in \mathbb{R}^2$ are constant vectors (see Fig. 2.1 in Chap. 2). Let $\gamma_i, i = 1, 2, \ldots, p$, denote the angles of the lines l_i such that

$$0 < \gamma_1 < \gamma_2 < \cdots < \gamma_p < 2\pi.$$

(A2)   There exist real-valued constant 2×2 matrices $A_1, A_2, \ldots, A_p$ defined by

$$A_i = \begin{bmatrix} \alpha_i & -\beta_i \\ \beta_i & \alpha_i \end{bmatrix} \text{ with } \beta_i > 0 \text{ for each } i = 1, 2, \ldots, p.$$

Meanwhile, for convenience throughout this section, we adopt the notations below.

(N1) $\theta_1 = (2\pi + \gamma_1) - \gamma_p, \theta_i = \gamma_i - \gamma_{i-1}, i = 2, 3, \ldots, p.$

(N2)   Let D_i denote the region situated between the straight lines l_{i-1} and l_i and defined in polar coordinates (r, ϕ), where $x_1 = r \cos \phi$, $x_2 = r \sin \phi$, as follows

$$D_1 = \{(r, \phi) \mid r \geq 0 \text{ and } \gamma_p < \phi \leq \gamma_1 + 2\pi\},$$

$$D_i = \{(r, \phi) \mid r \geq 0 \text{ and } \gamma_{i-1} < \phi \leq \gamma_i\}, \ i = 2, 3, \ldots, p.$$

Now, we define a function f such that $f(x) = A_i x$ for $x \in D_i, i = 1, 2, \ldots, p$, and consider a differential equation of the form

$$\frac{dx}{dt} = f(x). \tag{3.1.1}$$

According to the definition of the regions D_i, one can see that the function f in system (3.1.1) has discontinuities on the straight lines $l_i, i = 1, 2, \ldots, p$ (see Fig. 2.1 in Chap. 2).

Remark 3.1.1 It follows from the assumptions $(A1)$ and $(A2)$ that

$$\langle \frac{\partial \Phi_i(x)}{\partial x}, f(x) \rangle \neq 0 \ \text{ for } x \in l_i, \ i = 1, 2, \ldots, p.$$

That is, the vector field is transversal at every point on l_i for each i.

Using the polar transformation, we can write (3.1.1) in the following form

$$\frac{dr}{d\phi} = g(r), \qquad\qquad (3.1.2)$$

where

$$g(r) = \begin{cases} \lambda_1 r, \text{ if } \phi \in (\gamma_p + 2k\pi, \gamma_1 + 2(k+1)\pi], \\ \lambda_i r, \text{ if } \phi \in (\gamma_{i-1} + 2k\pi, \gamma_i + 2k\pi], \ i = 2, 3, \ldots, p, \end{cases}$$

with $\lambda_i = \dfrac{\alpha_i}{\beta_i}, i = 1, 2, \ldots, p$, and $k \in \mathbb{Z}$. Since Eq. (3.1.2) is 2π-periodic, it will be enough to consider just the section $\phi \in [0, 2\pi]$. Thus, the function g in (3.1.2) can be defined shortly as $g(r) = \lambda_i r$ if $(r, \phi) \in D_i$. Clearly, this function has discontinuities when $\phi = \gamma_i, i = 1, 2, \ldots, p$.

The solution $r(\phi, r_0)$ of (3.1.2) starting at the point $(0, r_0)$ has the form

$$r(\phi, r_0) = \begin{cases} \exp(\lambda_1 \phi) r_0, & \text{if } 0 \leq \phi \leq \gamma_1, \\ \exp(\lambda_1 \gamma_1 + \lambda_2 \theta_2 + \cdots + \lambda_i(\phi - \gamma_{i-1})) r_0, & \text{if } \gamma_{i-1} < \phi \leq \gamma_i, \\ \exp(\lambda_1 (\phi - (\gamma_p - \gamma_1)) + \sum_{i=2}^{p} \lambda_i \theta_i) r_0, & \text{if } \gamma_p < \phi \leq 2\pi, \end{cases}$$

where $i = 2, 3, \ldots, p$.

If we construct the Poincaré return map $r(2\pi, r_0)$ on the positive half-axis Ox_1, we get $r(2\pi, r_0) = \exp(\sum_{i=1}^{p} \lambda_i \theta_i) r_0$.

It is well-known that the origin is said to be a *center* if there exists a neighborhood of the origin where all trajectories are cycles surrounding the origin. Besides, if we can find a neighborhood of the origin such that all trajectories starting in it spiral to the origin as $t \to \infty$ ($t \to -\infty$), we call the origin as a stable (unstable) *focus*.

Let us denote $q = \exp(\sum_{i=1}^{p} \lambda_i \theta_i)$. Since $r(2\pi, r_0) = q r_0$, we obtain the following theorem for the nonperturbed system.

Theorem 3.1.1 *If*
(i) $q = 1$, then the origin is a center and all solutions are periodic with period
$$T = \sum_{i=1}^{p} \frac{\theta_i}{\beta_i};$$

(ii) $q < 1$, *then the origin is a stable focus;*
(iii) $q > 1$, *then the origin is an unstable focus of (3.1.1).*

3.1.3 The Perturbed System

Let $\Omega \subset \mathbb{R}^2$ be a domain in the neighborhood of the origin. When a function $u(x)$ is big-oh, or small-oh, of $v(x)$ as $x \to 0$, we shall denote it by $u(x) = O(v(x))$, or $u(x) = o(v(x))$, respectively. The following is the list of conditions assumed for this section.

(P1) Let $\{c_i\}_{i=1}^{p}$ be a set of curves in Ω which start at the origin and are determined by the equations $\tilde{\Phi}_i(y) = 0$, $\tilde{\Phi}_i(y) = \langle a^i, y \rangle + \tau_i(y)$, $i = 1, 2, \ldots, p$, where $\tau_i(y) = o(\|y\|)$, and for each i, the constant vectors a^i are the same as described in $(A1)$.

We split the domain Ω into p-subdomains, which will be called $\tilde{D}_i$ and formulated soon, by means of the curves c_i, $i = 1, 2, \ldots, p$. We assume without loss of generality that $\gamma_i \neq \frac{\pi}{2}j$, $j = 1, 3$. Then, for sufficiently small r, equation of the curve c_i can be written in polar coordinates as follows [6]:

$$c_i : \phi = \gamma_i + \psi_i(r, \phi), \quad i = 1, 2, \ldots, p, \tag{3.1.3}$$

where ψ_i is a 2π-periodic function in ϕ, continuously differentiable, and moreover $\psi_i = O(r)$. Using this discussion which makes use of polar transformation, we get

$$\tilde{D}_1 = \{(r, \phi) \mid r \geq 0 \text{ and } \gamma_p + \psi_p(r, \phi) < \phi \leq \gamma_1 + 2\pi + \psi_1(r, \phi)\},$$

$$\tilde{D}_i = \{(r, \phi) \mid r \geq 0 \text{ and } \gamma_{i-1} + \psi_{i-1}(r, \phi) < \phi \leq \gamma_i + \psi_i(r, \phi)\},$$
$$i = 2, 3, \ldots, p.$$

Let ε be a positive number and $N_\varepsilon(\tilde{D}_i)$ denote the ε-neighborhoods of the regions $\tilde{D}_i$, $i = 1, 2, \ldots, p$.

(P2) Let f_i be a function defined on $N_\varepsilon(\tilde{D}_i)$ and $f_i \in C^{(2)}(N_\varepsilon(\tilde{D}_i))$ for each $i = 1, 2, \ldots, p$.
(P3) $\tau_i \in C^{(2)}(N_\varepsilon(\tilde{D}_i))$, $i = 1, 2, \ldots, p$.
(P4) $f_i(y) = o(\|y\|)$, $i = 1, 2, \ldots, p$.

We shall consider the function $\tilde{f}(y) = A_i y + f_i(y)$ for $y \in \tilde{D}_i$, where the matrix A_i is as described in the assumption $(A2)$. On Ω, we now study the following differential equation associated with (3.1.1)

$$\frac{dy}{dt} = \tilde{f}(y), \tag{3.1.4}$$

where the function $\tilde{f}(y)$ has discontinuities on the curves $c_i, i = 1, 2, \ldots, p$.

If Ω is sufficiently small, then conditions $(A1)$ and $(P1)$ imply that the curves c_i intersect each other only at the origin, and none of them can intersect itself and

$$\langle \frac{\partial \tilde{\Phi}_i(y)}{\partial y}, \tilde{f}(y) \rangle \neq 0 \text{ for } y \in c_i, i = 1, 2, \ldots, p.$$

Further, for system (3.1.4) if a solution which starts sufficiently close to the origin on a curve c_i with fixed i, then conditions mentioned above imply the continuation of the solution to the curve c_{i+1} or c_{i-1} depending on the direction of the time.

We can utilize polar coordinates and assume that system (3.1.4) transforms into an equivalent system of the form

$$\frac{dr}{d\phi} = \tilde{g}(r, \phi), \tag{3.1.5}$$

where $\tilde{g}(r, \phi) = \lambda_i r + P_i(r, \phi)$ for $(r, \phi) \in \tilde{D}_i$. The function P_i is 2π-periodic in ϕ, continuously differentiable and $P_i = o(r), i = 1, 2, \ldots, p$.

From the construction, we see that system (3.1.5) is a differential equation with discontinuous right-hand side and the discontinuities occur on the curves c_i, $i = 1, 2, \ldots, p$. In almost every area of differential equations, it is common to reduce a given equation into an equivalent form by proper methods. From this point of view, we shall use the B-equivalence method [46, 54] which plays the role of a bridge in the passage from differential equations with discontinuous right-hand side to impulsive differential equations.

To reduce the system (3.1.5) with discontinuous vector fields into an impulsive differential equation, we redefine the function $\tilde{g}$ in the neighborhoods of the straight lines l_i, which contain the curve c_i. That is to say, we construct a new function g_N which is continuous everywhere except possibly at the points $(r, \phi) \in l_i$. The redefinition will be made at the points which lie between l_i and c_i and belong to the regions D_i or D_{i+1} for each i. Therefore, the construction is performed with minimal possible changes corresponding to the B-equivalence method, which is the main instrument of our investigation.

It is clear from the context that if $i = p$, then $D_{p+1} = D_1$. Using the argument above, we realize the following reconstruction of the domain. We consider the subregions of D_i and D_{i+1}, which are placed between the straight line l_i and the curve c_i. We refer to the subregions $D_i \cap \tilde{D}_{i+1}$ (horizontally shaded regions in Fig. 2.2) and $D_{i+1} \cap \tilde{D}_i$ (vertically shaded regions in Fig. 2.2 in Chap. 2) for all i. We extend the function $\tilde{g}$ from the region $D_i \cap \tilde{D}_{i+1}$ to D_i and from $D_{i+1} \cap \tilde{D}_i$ to D_{i+1} so that the new function g_N and its partial derivatives become continuous up to the angle $\phi = \gamma_i, i = 1, 2, \ldots, p$. According to all these discussions made for the definition of g_N, we conclude that $g_N(r, \phi) = \lambda_i r + P_i(r, \phi)$ for $(r, \phi) \in D_i$. Now, we consider the following differential equation

$$\frac{dr}{d\phi} = g_N(r, \phi). \tag{3.1.6}$$

Fix $i \in \{1, 2, \ldots, p\}$, and consider a neighborhood of l_i based on the description above. We need to analyze the following three cases:

I. Assume that the point $(r, \gamma_i) \in \tilde{D}_{i+1}$. Let $r^0(\phi) = r(\phi, \gamma_i, \rho)$ be a solution of (3.1.5) satisfying $r^0(\gamma_i) = \rho$ and ξ_i be the angle where this solution crosses the curve c_i. We denote a solution of (3.1.6) by $r^1(\phi) = r(\phi, \xi_i, r^0(\xi_i))$, $r^1(\xi_i) = r^0(\xi_i)$, on the interval $[\xi_i, \gamma_i]$. By the variation of constant formula, these solutions have the form

$$r^0(\phi) = \exp(\lambda_{i+1}(\phi - \gamma_i))\rho + \int_{\gamma_i}^{\phi} \exp(\lambda_{i+1}(\phi - s))P_{i+1}(r^0(s), s)ds,$$

$$r^1(\phi) = \exp(\lambda_i(\phi - \xi_i))r^0(\xi_i) + \int_{\xi_i}^{\phi} \exp(\lambda_i(\phi - s))P_i(r^1(s), s)ds.$$

Now, we define a mapping I_i on the line $\phi = \gamma_i$ into itself as follows:

$$\begin{aligned}
I_i(\rho) = r^1(\gamma_i) - \rho &= (\exp((\lambda_i - \lambda_{i+1})(\gamma_i - \xi_i)) - 1)\rho \\
&+ \exp(\lambda_i(\gamma_i - \xi_i))\int_{\gamma_i}^{\xi_i} \exp(\lambda_{i+1}(\xi_i - s))P_{i+1}ds \\
&+ \int_{\xi_i}^{\gamma_i} \exp(\lambda_i(\gamma_i - s))P_i ds.
\end{aligned}$$

II. If the point $(r, \gamma_i) \in \tilde{D}_i$, one can find I_i in a similar manner:

$$\begin{aligned}
I_i(\rho) &= (\exp((\lambda_i - \lambda_{i+1})(\xi_i - \gamma_i)) - 1)\rho \\
&+ \exp(\lambda_{i+1}(\gamma_i - \xi_i))\int_{\gamma_i}^{\xi_i} \exp(\lambda_i(\xi_i - s))P_i ds \\
&+ \int_{\xi_i}^{\gamma_i} \exp(\lambda_{i+1}(\gamma_i - s))P_{i+1}ds.
\end{aligned}$$

III. If $(r, \gamma_i) \in c_i$, then $I_i(\rho) = 0$.

Results from [6] imply that the functions I_i, $i = 1, 2, \ldots, p$, are continuously differentiable and the Eq. (3.1.3) leads us to $I_i = o(\rho)$.

Hereby, we construct the following impulsive differential equation

$$\begin{aligned}
\frac{d\rho}{d\phi} &= g_N(\rho, \phi), \qquad \phi \neq \gamma_i, \\
\Delta\rho|\phi &= \gamma_i = I_i(\rho).
\end{aligned} \tag{3.1.7}$$

Let $r(\phi, r_0)$ be a solution of (3.1.5), $r(0, r_0) = r_0$, and ξ_i be the meeting angle of this solution with the curve c_i. Denote by $(\xi_i, .\gamma_i]$ the interval $(\xi_i, \gamma_i]$ whenever $\xi_i \leq \gamma_i$ and $[\gamma_i, \xi_i)$ if $\gamma_i < \xi_i$.

Definition 3.1.1 We shall say that systems (3.1.5) and (3.1.7) are B-equivalent in Ω if for every solution $r(\phi, r_0)$ of (3.1.5) whose trajectory is in Ω for all $\phi \in [0, 2\pi]$ there exists a solution $\rho(\phi, r_0)$ of (3.1.7) which satisfies the relation

$$r(\phi, r_0) = \rho(\phi, r_0), \quad \phi \in [0, 2\pi] \setminus \bigcup_{i=1}^{p} (\xi_i, .\gamma_i], \tag{3.1.8}$$

and, conversely, for every solution $\rho(\phi, r_0)$ of (3.1.7) whose trajectory is in Ω, there exists a solution $r(\phi, r_0)$ of (3.1.5) which satisfies (3.1.8).

From the discussion above and the construction of the impulsive system (3.1.7) with impulse actions at fixed angles, it follows that for sufficiently small Ω, solution $r(\phi, r_0)$ of (3.1.5) whose trajectory is in Ω for all $\phi \in [0, 2\pi]$ takes the same values with the exception of the oriented intervals $(\xi_i, .\gamma_i]$ as the solution $\rho(\phi, r_0)$, $\rho(0, r_0) = r_0$, of (3.1.7). Hence, systems (3.1.5) and (3.1.7) are B-equivalent in the sense of the Definition 3.1.1. Moreover, solutions of (3.1.5) exist in the neighborhood Ω, and they are continuous and have discontinuities in the derivative on the curves c_i. Correspondingly, a solution of system (3.1.4) for any initial value is continuous, continuously differentiable except possibly at the moments when the trajectories intersect the curves c_i, and it is unique.

Theorem 3.1.2 *Suppose* $(A1) - (A2)$, $(P1) - (P4)$ *are satisfied and* $q < 1$ $(q > 1)$. *Then the origin is a stable (unstable) focus of (3.1.4).*

Proof Let $r(\phi, r_0)$ be the solution of (3.1.5) with $r(0, r_0) = 0$ and $\rho(\phi, r_0)$, $\rho(0, r_0) = r_0$, be the solution of (3.1.7). For the sake of simplicity, we shall use the notations $P_i = P_i(\rho(s, r_0), s)$ and $I_i = I_i(\rho(\gamma_i, r_0))$, $i = 1, 2, \ldots, p$.

On the interval $\phi \in [0, \gamma_1]$, we have

$$\rho(\phi, r_0) = \exp(\lambda_1 \phi) r_0 + \int_0^{\phi} \exp(\lambda_1(\phi - s)) P_1 ds.$$

For any i, $2 \leq i \leq p$, the solution $\rho(\phi, r_0)$ of (3.1.7) on $(\gamma_{i-1}, \gamma_i]$ is given by

$$\begin{aligned}
\rho(\phi, r_0) &= \exp\left(\lambda_i(\phi - \gamma_{i-1}) + \lambda_{i-1}\theta_{i-1} + \cdots + \lambda_2\theta_2 + \lambda_1\gamma_1\right) r_0 \\
&\quad + \exp\left(\lambda_i(\phi - \gamma_{i-1}) + \cdots + \lambda_2\theta_2 + \lambda_1\gamma_1\right) \int_0^{\gamma_1} \exp(-\lambda_1 s) P_1 ds \\
&\quad + \sum_{k=2}^{i-1} \exp\left(\lambda_i(\phi - \gamma_{i-1}) + \cdots + \lambda_{k+1}\theta_{k+1} + \lambda_k\gamma_k\right) \int_{\gamma_{k-1}}^{\gamma_k} \exp(-\lambda_k s) P_k ds \\
&\quad + \int_{\gamma_{i-1}}^{\phi} \exp\left(\lambda_i(\phi - s)\right) P_i ds \\
&\quad + \sum_{k=2}^{i} \exp\left(\lambda_i(\phi - \gamma_{i-1}) + \lambda_{i-1}\theta_{i-1} + \cdots + \lambda_k\theta_k\right) I_{k-1}.
\end{aligned}$$

For $\phi \in (\gamma_p, 2\pi]$, system (3.1.7) admits the solution

$$\rho(\phi, r_0) = \exp\left(\lambda_1(\phi - \gamma_p)\right)\left(\rho(\gamma_p, r_0) + I_p\right) + \int_{\gamma_p}^{\phi} \exp\left(\lambda_1(\phi - s)\right) P_1 ds.$$

Using the differentiable dependence of solutions of impulse systems on parameters [55] and the results from [6], we can conclude that the solution $\rho(\phi, r_0)$ is differentiable in r_0 and $\frac{\partial \rho(\phi, r_0)}{\partial r_0}|_{(\phi, r_0)=(2\pi, 0)} = q$. Since systems (3.1.5) and (3.1.7), correspondingly (3.1.4) and (3.1.7), are B-equivalent, we derive

$$\frac{\partial r(\phi, r_0)}{\partial r_0}|_{(\phi, r_0)=(2\pi, 0)} = q,$$

which completes the proof. $\square$

3.1.4 The Focus-Center Problem

If $q = 1$, then we have the critical case and the origin is either a focus or a center for system (3.1.4). In what follows, we solve this problem of distinguishing between the focus and the center.

We assume that f_i and τ_i, $i = 1, 2, \ldots, p$, are analytic functions in $N_\varepsilon(\tilde{D}_i)$. Then, for sufficiently small ρ, the solution $\rho(\phi, r_0)$ of (3.1.7) satisfying $\rho(0, r_0) = r_0$ has the expansion [54]

$$\rho(\phi, r_0) = \sum_{j=0}^{\infty} \rho_j(\phi) r_0^j, \tag{3.1.9}$$

for all $\phi \in [0, 2\pi]$. From the expansion (3.1.9), it can be easily seen that $\rho_1(0) = 1$, $\rho_i(0) = 0$ for all $i = 0, 2, 3, 4, \ldots$, and $\rho_0(\phi) = 0$. The coefficient $\rho_1(\phi)$ with $\rho_1(0) = 1$ is the solution of the system

$$\frac{d\rho_1}{d\phi} = g(\rho_1),$$

where g is the function defined in system (3.1.2). It is clear that $\rho_1(2\pi) = q = 1$. We use the notation $k_j = \rho_j(2\pi)$, $j = 2, 3, \ldots$. For the solution $\rho(\phi, r_0)$ of (3.1.7), we construct the Poincaré return map as follows:

$$\rho(2\pi, r_0) = qr_0 + \sum_{j=2}^{\infty} k_j r_0^j.$$

In the critical case, the sign of the first nonzero element of the sequence k_j determines what type of a singular point the origin is. Moreover, for all $i = 1, 2, \ldots, p$, we have

$$P_i(\rho, \phi) = \sum_{j=2}^{\infty} P_{ij}(\phi)\rho^j, \tag{3.1.10}$$

and

$$I_i(\rho) = \sum_{j=2}^{\infty} I_{ij}\rho^j. \tag{3.1.11}$$

The existence of the expansions (3.1.10) and (3.1.11) has been proved in [54]. By means of (3.1.10) and (3.1.11), one can derive that the coefficients $\rho_j(\phi)$ with $\rho_j(0) = 0$, $j = 2, 3, \ldots$, are solutions of the following impulsive system

$$\begin{aligned} \frac{d\rho_j}{d\phi} &= h(\rho_j, \phi), \qquad \phi \neq \gamma_i, \\ \Delta\rho_j|_{\phi = \gamma_i} &= W_{ij}, \end{aligned} \tag{3.1.12}$$

where $h(\rho_j, \phi) = \lambda_i \rho_j + Q_{ij}(\phi)$ if $(\rho_j, \phi) \in D_i$, $i = 1, 2, \ldots, p$. From the differential part of (3.1.7) and the expansion (3.1.10), one can evaluate for any i, $1 \leq i \leq p$,

$$Q_{i2}(\phi) = P_{i2}(\phi)\rho_1^2(\phi), \quad Q_{i3}(\phi) = 2P_{i2}(\phi)\rho_1(\phi)\rho_2(\phi) + P_{i3}(\phi)\rho_1^3(\phi)$$

and $Q_{ij}(\phi)$, for $j = 4, 5, \ldots$, can be determined similarly. Further, the constants W_{ij} in (3.1.12) can be found from the impulsive part of (3.1.7) and the expansion (3.1.11). For instance,

$$W_{i2} = I_{i2}\rho_1^2(\gamma_i), \quad W_{i3} = 2I_{i2}\rho_1(\gamma_i)\rho_2(\gamma_i) + I_{i3}\rho_1^3(\gamma_i),$$

and W_{ij} can be evaluated, for $j = 4, 5, \ldots$, in the same manner.

As $k_j = \rho_j(2\pi)$, by solving the system (3.1.12), one can evaluate k_j, $j = 2, 3, \ldots$, which are the coefficients in the expansion of the Poincaré return map $\rho(2\pi, r_0)$:

$$\begin{aligned} k_j &= \int_0^{\gamma_1} \exp(-\lambda_1 s)Q_{1j}ds + \int_{\gamma_p}^{2\pi} \exp\left(\lambda_1(2\pi - s)\right) Q_{1j}ds + \\ &\quad \sum_{i=2}^{p} \exp\left(\lambda_1(2\pi - \gamma_p) + \cdots + \lambda_{i+1}\theta_{i+1} + \lambda_i\gamma_i\right) \int_{\gamma_{i-1}}^{\gamma_i} \exp(-\lambda_i s)Q_{ij}ds + \\ &\quad \sum_{i=2}^{p} \exp\left(\lambda_1(2\pi - \gamma_p) + \lambda_p\theta_p + \cdots + \lambda_i\theta_i\right) W_{i-1,j} + \exp\left(\lambda_1(2\pi - \gamma_p)\right) W_{pj}. \end{aligned} \tag{3.1.13}$$

From the expansion of $\rho(2\pi, r_0)$ and (3.1.13), it immediately follows that the following assertion is valid.

Lemma 3.1.1 *Let* $q = 1$ *and the first nonzero element of the sequence* k_j, $j = 2, 3, \ldots$, *be negative (positive). Then the origin is a stable (unstable) focus of (3.1.7). If* $k_j = 0$ *for all* $j \geq 2$, *then the origin is a center for system (3.1.7).*

Since systems (3.1.5) and (3.1.7), correspondingly (3.1.4) and (3.1.7), are *B*-equivalent, we have proved the following theorem.

Theorem 3.1.3 *Let* $q = 1$ *and the first nonzero element of the sequence* k_j, $j = 2, 3, \ldots$, *be negative (positive). Then the origin is a stable (unstable) focus of (3.1.4). If* $k_j = 0$ *for all* $j \geq 2$, *then the origin is a center for system (3.1.4).*

3.1.5 Bifurcation of Periodic Solutions

In this section, we first introduce the system

$$\frac{dz}{dt} = \hat{f}(z, \mu) \tag{3.1.14}$$

where $\hat{f}(z, \mu) = A_i z + f_i(z) + \mu F_i(z, \mu)$ for $z \in \tilde{D}_i(\mu) \subset \mathbb{R}^2$ for analysis, and then, we will describe it in detail with the help of the following assumptions:

(H1) Let $\{c_i(\mu)\}_{i=1}^p$ be a collection of curves in Ω which start at the origin and are given by the equations $\langle a^i, z \rangle + \tau_i(z) + \mu \kappa_i(z, \mu) = 0$, $i = 1, 2, \ldots, p$.

(H2) Let $\{l_i(\mu)\}_{i=1}^p$ be a union of half-lines which start at the origin and are defined by $\langle a^i + \mu \dfrac{\partial \kappa_i(0, \mu)}{\partial z}, z \rangle = 0$, $i = 1, 2, \ldots, p$. Denote by $\gamma_i(\mu)$ the angles of the lines $l_i(\mu)$, $i = 1, 2, \ldots, p$.

Similar to the construction of the regions D_i and $\tilde{D}_i$, we set for $\mu \in (-\mu_0, \mu_0)$ and $i = 2, 3, \ldots, p$:

$$\tilde{D}_1(\mu) = \{(r, \phi, \mu) \mid r \geq 0, \ \gamma_p(\mu) + \Psi_p < \phi \leq \gamma_1(\mu) + 2\pi + \Psi_1\},$$

$$\tilde{D}_i(\mu) = \{(r, \phi, \mu) \mid r \geq 0, \ \gamma_{i-1}(\mu) + \Psi_{i-1} < \phi \leq \gamma_i(\mu) + \Psi_i\},$$

$$D_1(\mu) = \{(r, \phi, \mu) \mid r \geq 0, \ \gamma_p(\mu) < \phi \leq \gamma_1(\mu) + 2\pi\},$$

$$D_i(\mu) = \{(r, \phi, \mu) \mid r \geq 0, \ \gamma_{i-1}(\mu) < \phi \leq \gamma_i(\mu)\},$$

where functions $\Psi_i = \Psi_i(r, \phi, \mu)$ are 2π-periodic in ϕ, continuously differentiable, $\Psi_i = O(r)$, $i = 1, 2, \ldots, p$, and they can be defined applying a similar technique used in the construction of Eq. (3.1.3).

(H3) $F_i: N_\varepsilon(\tilde{D}_i(\mu)) \times (-\mu_0, \mu_0) \to \mathbb{R}^2$ and κ_i are analytical functions both in z and in μ in the ε-neighborhood of their domains.

(H4) $F_i(0, \mu) = 0$ and $\kappa_i(0, \mu) = 0$ hold uniformly for each i and $\mu \in (-\mu_0, \mu_0)$.
(H5) The matrices A_i, the functions f_i, τ_i, and the constant vectors a^i correspond
      to the ones described in systems (3.1.1) and (3.1.4).

Besides the system (3.1.14), we need the equation

$$\frac{dz}{dt} = \hat{f}_z(0, \mu)z, \tag{3.1.15}$$

where $\hat{f}_z(0, \mu) = A_i + \mu \dfrac{\partial F_i(0, \mu)}{\partial z}$ whenever $z \in D_i(\mu)$.

In polar coordinates, system (3.1.14) reduces to

$$\frac{dr}{d\phi} = \hat{g}(r, \phi, \mu), \tag{3.1.16}$$

where $\hat{g}(r, \phi, \mu) = \lambda_i(\mu)r + P_i(r, \phi, \mu)$ if $(r, \phi, \mu) \in \tilde{D}_i(\mu)$.

Let the following impulse system

$$\begin{aligned}
\frac{d\rho}{d\phi} &= \hat{g}_N(\rho, \phi, \mu), \qquad \phi \neq \gamma_i(\mu), \\
\Delta\rho|_\phi &= \gamma_i(\mu) = I_i(\rho, \mu)
\end{aligned} \tag{3.1.17}$$

be B-equivalent to (3.1.16), where $\hat{g}_N$ stands for the extension of $\hat{g}$ as we described
in Sect. 3.1.3. That is, $\hat{g}_N(\rho, \phi, \mu) = \lambda_i(\mu)\rho + P_i(\rho, \phi, \mu)$ for $(\rho, \phi, \mu) \in D_i(\mu)$.
We know that the function $\hat{g}_N$ and its partial derivatives become continuous up to the
angle $\phi = \gamma_i(\mu)$ for $i = 1, 2, \ldots, p$. The function $I_i(\rho, \mu)$, for each $i = 1, 2, \ldots, p$
can be defined in the same way as done for $I_i(\rho)$.

Using a similar argument as in (3.1.1), we can obtain for system (3.1.15) that

$$q(\mu) = \exp\left(\sum_{i=1}^{p} \lambda_i(\mu)\theta_i(\mu)\right).$$

The last expression plays an important rule to establish the theorem on the bifurcation of periodic solutions as stated below. To prove the assertion, we apply the
technique of paper [6].

Theorem 3.1.4 *Let $q(0) = 1$, $q'(0) \neq 0$ and the origin be a focus for (3.1.4).
Then, for sufficiently small r_0, there exists a unique continuous function $\mu = \delta(r_0)$,
$\delta(0) = 0$, such that the solution $r(\phi, r_0, \delta(r_0))$ of (3.1.16) is periodic with period
2π. Moreover, the closed trajectory is stable (unstable) if the origin of (3.1.4) is a
stable (unstable) focus. The period of the corresponding periodic solution of (3.1.14)
is $T = \sum_{i=1}^{p} \frac{\theta_i}{\beta_i} + o(|\mu|)$.*

Proof Let $\rho(\phi, r_0, \mu)$ be the solution of (3.1.17) such that $\rho(0, r_0, \mu) = r_0$. To exclude the trivial solution, we consider $r_0 > 0$. The theorem of analyticity of solutions [54] implies that

$$\rho(2\pi, r_0, \mu) = \sum_{j=1}^{\infty} k_j(\mu) r_0^j,$$

where $k_j(\mu) = \sum_{i=0}^{\infty} k_{ji} \mu^i$. Since $k_1(\mu) = q(\mu)$, we have by the hypotheses of the theorem that $k_{10} = q(0) = 1$ and $k_{11} = q'(0) \neq 0$. For the existence of a periodic solution, we require that $\rho(2\pi, r_0, \mu) = r_0$. Now, we define $\mathscr{F}(r_0, \mu) = \rho(2\pi, r_0, \mu) - r_0$. Then, it can be derived that

$$\mathscr{F}(r_0, \mu) = q'(0)\mu r_0 + \sum_{j=2}^{\infty} k_{j0} r_0^j + \sum_{i+j\geq 3} k_{ji} \mu^i r_0^j,$$

where $i, j \in \mathbb{N}$ in the second summation. We call $\mathscr{F}(r_0, \mu) = 0$ as the bifurcation equation. If we cancel by r_0, we obtain the equation

$$\mathscr{H}(r_0, \mu) = 0, \tag{3.1.18}$$

where

$$\mathscr{H}(r_0, \mu) = q'(0)\mu + \sum_{j=2}^{\infty} k_{j0} r_0^{j-1} + \sum_{i+j\geq 2} k_{j+1,i} \mu^i r_0^j.$$

In the second summation of the last equation, we have $i = 1, 2\ldots,$ and $j = 0, 1, \ldots$. Since $\mathscr{H}(0,0) = 0$ and $\dfrac{\partial \mathscr{H}(0,0)}{\partial \mu} = q'(0) \neq 0$, one can say by the implicit function theorem that for sufficiently small r_0, there exists a function $\mu = \delta(r_0)$ such that $\rho(\phi, r_0, \delta(r_0))$ is a periodic solution.

We assume without loss of generality that $k_{j0} = 0$ for $j = 2, 3, \ldots, l-1$ and $k_{l0} \neq 0$. Then, we can obtain from (3.1.18) that

$$\delta(r_0) = -\frac{k_{l0}}{q'(0)} r_0^{l-1} + \sum_{i=l}^{\infty} \delta_i r_0^i. \tag{3.1.19}$$

If we analyze the Eq. (3.1.19), we can conclude that the bifurcation of periodic solutions exists if a stable (unstable) focus for $\mu = 0$ becomes unstable (stable) for $\mu \neq 0$.

Let $\rho(\phi) = \rho(\phi, \bar{r}_0, \bar{\mu})$ be a periodic solution of (3.1.17). This periodic solution is a limit cycle if $\dfrac{\partial \mathscr{F}(\bar{r}_0, \bar{\mu})}{\partial r_0} < 0$. Assuming that the first nonzero element k_{l0} of the sequence k_{j0}, $j \geq 2$, is negative and using (3.1.19), we get

$$\frac{\partial \mathscr{F}(\bar{r}_0, \bar{\mu})}{\partial r_0} = (l-1)k_{l0}\bar{r}_0{}^{l-1} + \mathscr{G}(r_0),$$

where $\mathscr{G}$ starts with a member whose order is not less than l. Thus, it implies that $\dfrac{\partial \mathscr{F}(\bar{r}_0, \bar{\mu})}{\partial r_0} < 0$.

Since (3.1.16) and (3.1.17) are B-equivalent systems, the proof is completed.

3.1.6 An Example

To be convenient, in the following example, we use the corresponding notations that are adopted above.

Example 3.1.1 Let $c_1(\mu)$ and $c_2(\mu)$ be the curves defined by $z_2 = \frac{1}{\sqrt{3}}z_1 + (1+\mu)z_1^3$, $z_1 > 0$ and $z_2 = \sqrt{3}z_1 + z_1^5 + \mu z_1^2$, $z_1 < 0$, respectively. We take

$$A_1 = \begin{bmatrix} -0.7 & -2 \\ 2 & -0.7 \end{bmatrix}, \quad f_1(z) = \begin{bmatrix} z_1\sqrt{z_1^2 + z_2^2} \\ z_2\sqrt{z_1^2 + z_2^2} \end{bmatrix}, \quad F_1(z, \mu) = \begin{bmatrix} z_1 \\ z_2 \end{bmatrix},$$

and

$$A_2 = \begin{bmatrix} 0.5 & -2 \\ 2 & 0.5 \end{bmatrix}, \quad f_2(z) = \begin{bmatrix} -2z_1\sqrt{z_1^2 + z_2^2} \\ -2z_2\sqrt{z_1^2 + z_2^2} \end{bmatrix}, \quad F_2(z, \mu) = \begin{bmatrix} -z_1 \\ -z_2 \end{bmatrix}.$$

After these preparations, we consider the system

$$\frac{dz}{dt} = \hat{f}(z, \mu) \tag{3.1.20}$$

where $\hat{f}(z, \mu) = A_i z + f_i(z) + \mu F_i(z, \mu)$ if $z \in \tilde{D}_i(\mu)$, $i = 1, 2$. Here, $\tilde{D}_1(\mu)$ denotes the region situated between the curves $c_1(\mu)$ and $c_2(\mu)$, which contains the fourth quadrant. $\tilde{D}_1(\mu)$ is the region between $c_1(\mu)$ and $c_2(\mu)$ containing the second quadrant.

Since $q = 1$, by Theorem 3.1.1, the origin is a center for the nonperturbed system

$$\frac{dx}{dt} = f(x)$$

where $f(x) = A_i x$ whenever $x \in D_i$, $i = 1, 2$ as shown in Fig. 3.1. Here, D_1 and D_2 are the regions between the half straight lines $l_1 : z_2 = \frac{1}{\sqrt{3}}z_1$, $z_1 > 0$ and $l_2 : z_2 = \sqrt{3}z_1$, $z_1 < 0$, which contain the fourth and second quadrants, respectively.

Fig. 3.1 The simulation
result showing the existence
of a center for the
nonperturbed system

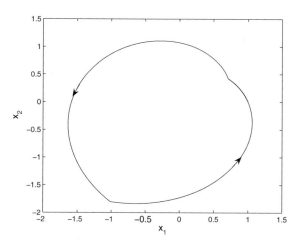

One can see that $l_1(\mu)$ $(l_2(\mu))$ coincides with l_1 (l_2). Hence, $\gamma_1 = \gamma_1(\mu) = \frac{\pi}{6}$ and
$\gamma_2 = \gamma_2(\mu) = \frac{4\pi}{3}$. Using the given information, we obtain

$$q(\mu) = \exp\left(-\frac{\pi}{6}\mu\right), \quad q(0) = 1, \quad q'(0) = -\frac{\pi}{6} \neq 0.$$

Moreover, for the associated system

$$\frac{dy}{dt} = \tilde{f}(y),$$

where $\tilde{f}(y) = A_i y + f_i(y)$ whenever $y \in \tilde{D}_i$, $i = 1, 2$, it follows from Theorem
3.1.3 that the origin is a stable focus as $k_2 < 0$ for the perturbed system (see Fig. 3.2).

Fig. 3.2 The simulation
result showing the existence
of a stable focus for the
perturbed system ($\mu = 0$)

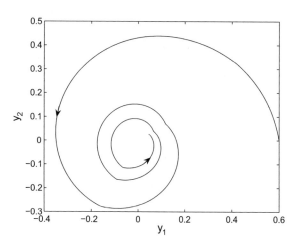

Fig. 3.3 The simulation result showing the existence of a limit cycle for system (3.1.20)

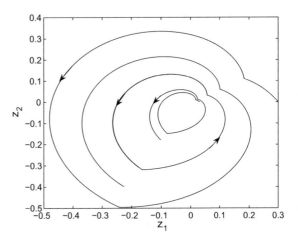

Here, $\tilde{D}_1$ and $\tilde{D}_2$ are the regions between the curves $c_1 : z_2 = \frac{1}{\sqrt{3}} z_1 + z_1^3$, $z_1 > 0$ and $c_2 : z_2 = \sqrt{3} z_1 + z_1^5$, $z_1 < 0$, which contain the fourth and second quadrants, respectively.

From Fig. 3.3, which is simulated for $\mu = -0.8$, we see that the trajectories approach to the periodic solution from interior and exterior. That is to say, system (3.1.20) has a limit cycle with period $\approx \pi$.

3.2 3D Filippov System

We study the behavior of solutions for a three-dimensional system of differential equations with discontinuous right-hand side in the neighborhood of the origin. Using B-equivalence of that system to an impulsive differential equation [46, 54], existence of a center manifold is proved, and then, a Hopf bifurcation theorem is provided for such equations in the critical case. The results are apparently obtained for the systems with dimensions greater than two for the first time. Finally, an appropriate example is given to illustrate our results.

3.2.1 Introduction

When we consider the bifurcations of a given type in a neighborhood of the origin, the center manifold theory appears as one of the most effective tools in the investigation. The study of center manifolds can be traced back to the works of Pliss [194] and Kelley [136]. When such manifolds exist, the investigation of local behaviors can be reduced to the study of the systems on the center manifolds. Any bifurcations which occur in the neighborhood of the origin on the center manifold are guaranteed to occur in the full nonlinear system as well. In particular, if a limit cycle exists on the center manifold, then it will also appear in the full system.

Physical phenomena are often modeled by discontinuous dynamical systems which switch between different vector fields in different modes. Filippov systems form a subclass of discontinuous systems described by differential equations with a discontinuous right-hand side [115]. Bifurcations in smooth systems are well understood, but little is known in discontinuous dynamical systems. In the last several decades, existence of nonsmooth dynamics in the real world has stimulated the study of bifurcation of periodic solutions in discontinuous systems [92, 127], [146, 147, 150, 173, 183, 184, 237]. Furthermore, Bautin and Leontovich [65] and Küpper et al. [150, 237] have considered Hopf bifurcation for planar Filippov systems with discontinuities on a single straight line. However, to the best of our knowledge, there have been no results considering bifurcation in three and more dimensions for equations with discontinuous vector fields.

In the previous section, Hopf bifurcation has been investigated for planar discontinuous dynamical systems. Based on the method of B-equivalence [46, 47, 54, 55] to impulsive differential equations and by using the projection on the center manifold, we extend the results in [6] to obtain qualitative properties for our three-dimensional system with discontinuous right-hand side. The present section deals with discontinuities on arbitrarily finite nonlinear surfaces. In fact, it is the advantage of the B-equivalence method that we can consider a system with nonlinear discontinuity sets.

The structure of the section is as follows. Section 3.2.2 describes the nonperturbed system and studies its qualitative properties. Section 3.2.3 is dedicated to the perturbed system and the notion of B-equivalent impulsive systems. The center manifold theory is given in Sect. 3.2.4. Our main results concerning the bifurcation of periodic solutions are formulated in Sect. 3.2.5. In the last section, we present an appropriate example to illustrate our findings.

3.2.2 The Nonperturbed System

Let $\mathbb{N}$ and $\mathbb{R}$ be the sets of natural and real numbers, respectively. Let $\mathbb{R}^n$, $n \in \mathbb{N}$, be the n-dimensional real space and $\langle x, y \rangle$ denote the scalar product for all vectors $x, y \in \mathbb{R}^n$. The norm of a vector $x \in \mathbb{R}^n$ is given by $\|x\| = \langle x, x \rangle^{\frac{1}{2}}$.

Also for the sake of brevity in the sequel, every angle for a point is considered with respect to the positive half-line of the first coordinate axis in $x_1 x_2$-plane. Moreover, it is important to note that we shall consider angle values only in the interval $[0, 2\pi]$ because of the periodicity.

Before introducing the nonperturbed system, we give the following assumptions and notations which will be needed throughout this section:

Fig. 3.4 Half-planes $\mathscr{P}_i$, $i = \overline{1, p}$, of discontinuities for the nonperturbed system (3.2.21)

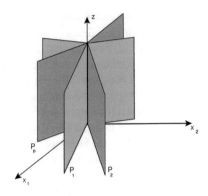

(A1)   Let $\{\mathscr{P}_i\}_{i=1}^{p}$, $p \geq 2$, $p \in \mathbb{N}$, be a set of half-planes starting at the z-axis, i.e., $\mathscr{P}_i = l_i \times \mathbb{R}$, where l_i are half-lines which start at the origin and are given by $\varphi_i(x) = 0$, $\varphi_i(x) = \langle a^i, x \rangle$, $x \in \mathbb{R}^2$, and $a^i = (a_1{}^i, a_2{}^i) \in \mathbb{R}^2$ are constant vectors (see Fig. 3.4). Let γ_i denote the angle of the line l_i for each $i = \overline{1, p}$ such that

$$0 < \gamma_1 < \gamma_2 < \cdots < \gamma_p < 2\pi \; ;$$

(A2)   There exist constant, real-valued 2×2 matrices A_i defined by $A_i = \begin{bmatrix} \alpha_i & -\beta_i \\ \beta_i & \alpha_i \end{bmatrix}$ where $\beta_i > 0$ and constants $b_i \in \mathbb{R}$, $i = \overline{1, p}$.

(N1) $\theta_1 = (2\pi + \gamma_1) - \gamma_p$ and $\theta_i = \gamma_i - \gamma_{i-1}$, $i = \overline{2, p}$;
(N2)   Let D_i denote the region situated between the planes $\mathscr{P}_{i-1}$ and $\mathscr{P}_i$ and defined in cylindrical coordinates (r, ϕ, z), where $x_1 = r \cos \phi$, $x_2 = r \sin \phi$ and $z = z$, by

$$D_1 = \{(r, \phi, z) \mid r \geq 0, \; \gamma_p < \phi \leq \gamma_1 + 2\pi, \; z \in \mathbb{R}\},$$

$$D_i = \{(r, \phi, z) \mid r \geq 0, \; \gamma_{i-1} < \phi \leq \gamma_i, \; z \in \mathbb{R}\}, \; i = \overline{2, p}.$$

Under the assumptions made above, we study in $\mathbb{R}^3$ the following nonperturbed system,

$$\begin{aligned} \frac{dx}{dt} &= F(x), \\ \frac{dz}{dt} &= f(z), \end{aligned} \qquad (3.2.21)$$

where $F(x) = A_i x$ and $f(z) = b_i z$ for $(x, z) \in D_i$, $i = \overline{1, p}$.

We note that the functions F and f in system (3.2.21) are discontinuous on the planes $\mathscr{P}_i$, $i = \overline{1, p}$.

Remark 3.2.1 It follows from the assumptions (A1) and (A2) that

$$\langle \frac{\partial \varphi_i(x)}{\partial x}, F(x) \rangle \neq 0 \ \text{ for } x \in l_i, \ i = \overline{1, p}.$$

That is, the vector field is transversal at every point on $\mathscr{P}_i$ for each i.

Since the results can be most conveniently stated in terms of cylindrical coordinates, we use the transformation $x_1 = r \cos \phi$, $x_2 = r \sin \phi$, $z = z$ so that system (3.2.21) reduces to

$$\begin{aligned} \frac{dr}{d\phi} &= G(r), \\ \frac{dz}{d\phi} &= g(z), \end{aligned} \tag{3.2.22}$$

where $G(r) = \lambda_i r$ and $g(z) = k_i z$ if $(r, \phi, z) \in D_i$, with $\lambda_i = \dfrac{\alpha_i}{\beta_i}$ and $k_i = \dfrac{b_i}{\beta_i}$, $i = \overline{1, p}$. We see that the functions G and g given in (3.2.22) have discontinuities when $\phi = \gamma_i$, $i = \overline{1, p}$.

The solution $(r(\phi, r_0), z(\phi, z_0))$ of (3.2.22) starting at the point $(0, r_0, z_0)$ is given by

$$r(\phi, r_0) = \begin{cases} \exp(\lambda_1 \phi) r_0, & \text{if } 0 \leq \phi \leq \gamma_1, \\ \exp\{\lambda_1 \gamma_1 + \lambda_2 \theta_2 + \cdots + \lambda_i(\phi - \gamma_{i-1})\} r_0, & \text{if } \gamma_{i-1} < \phi \leq \gamma_i, \\ \exp\{\lambda_1[\phi - (\gamma_p - \gamma_1)] + \sum_{i=2}^{p} \lambda_i \theta_i\} r_0, & \text{if } \gamma_p < \phi \leq 2\pi, \end{cases}$$

$$z(\phi, z_0) = \begin{cases} \exp(k_1 \phi) z_0, & \text{if } 0 \leq \phi \leq \gamma_1, \\ \exp\{k_1 \gamma_1 + k_2 \theta_2 + \cdots + k_i(\phi - \gamma_{i-1})\} z_0, & \text{if } \gamma_{i-1} < \phi \leq \gamma_i, \\ \exp\{k_1[\phi - (\gamma_p - \gamma_1)] + \sum_{i=2}^{p} k_i \theta_i\} z_0, & \text{if } \gamma_p < \phi \leq 2\pi, \end{cases}$$

for $i = 2, 3, \ldots, p$.

Now, we define a section $\mathsf{P} = \{(x_1, x_2, z) \mid x_2 = 0, x_1 > 0, z \in \mathbb{R}\}$. Constructing the Poincaré return map on P, we find that

$$(r(2\pi, r_0), z(2\pi, z_0)) = \left(\exp\left(\sum_{i=1}^{p} \lambda_i \theta_i\right) r_0, \exp\left(\sum_{i=1}^{p} k_i \theta_i\right) z_0 \right).$$

Let us denote

$$q_1 = \exp\left(\sum_{i=1}^{p} \lambda_i \theta_i\right), \tag{3.2.23}$$

$$q_2 = \exp\left(\sum_{i=1}^{p} k_i \theta_i\right). \tag{3.2.24}$$

Since $r(2\pi, r_0) = q_1 r_0$, $z(2\pi, z_0) = q_2 z_0$, we can establish the following assertions.

Lemma 3.2.1 *Assume that $q_1 = 1$. If*

(i) $q_2 = 1$, *then all solutions are periodic with period $T = \sum_{i=1}^{p} \frac{\theta_i}{\beta_i}$, i.e., $\mathbb{R}^3$ is a center manifold;*

(ii) $q_2 < 1$, *then a solution that starts to its motion on $x_1 x_2$-plane is T-periodic and all other solutions lie on the surface of a cylinder and they move toward the $x_1 x_2$-plane, i.e., $x_1 x_2$-plane is a center manifold and z-axis is a stable manifold;*

(iii) $q_2 > 1$, *then a solution that starts to its motion on $x_1 x_2$-plane is T-periodic and all other solutions lie on the surface of a cylinder and they move away from the origin, i.e., $x_1 x_2$-plane is a center manifold and z-axis is an unstable manifold.*

Lemma 3.2.2 *Assume that $q_1 < 1$. If*

(i) $q_2 = 1$, *then a solution that starts to its motion on z-axis is T-periodic and all other solutions will approach to z-axis, i.e., $x_1 x_2$-plane is a stable manifold and z-axis is a center manifold;*

(ii) $q_2 < 1$, *all solutions will spiral toward the origin, i.e., the origin is asymptotically stable;*

(iii) $q_2 > 1$, *a solution that starts to its motion on $x_1 x_2$-plane spirals toward the origin and a solution initiating on z-axis will move away from the origin, i.e., $x_1 x_2$-plane is a stable manifold and z-axis is a center manifold.*

Lemma 3.2.3 *Assume that $q_1 > 1$. If*

(i) $q_2 = 1$, *then a solution that starts to its motion on z-axis is T-periodic and all other solutions move away from the z-axis, i.e., $x_1 x_2$-plane is an unstable manifold and z-axis is a center manifold;*

(ii) $q_2 < 1$, *a solution that starts to its motion on $x_1 x_2$-plane moves away from the origin and a solution initiating on z-axis spirals toward the origin, i.e., $x_1 x_2$-plane is an unstable manifold and z-axis is a stable manifold;*

(iii) $q_2 > 1$, *all solutions move away from the origin, i.e., the origin is unstable.*

Remark 3.2.2 From now on, we assume that $q_1 = 1$ and $q_2 < 1$. In other words, $x_1 x_2$-plane is a center manifold and z-axis is a stable manifold.

3.2.3   The Perturbed System

Let $\Omega \subset \mathbb{R}^3$ be a domain in the neighborhood of the origin. The following conditions are assumed to hold throughout this section.

(P1)   Let $\{\mathscr{S}_i\}_{i=1}^p$, $p \geq 2$, be a set of cylindrical surfaces which start at the z-axis, i.e., $\mathscr{S}_i = c_i \times \mathbb{R}$, where c_i are curves starting at the origin and determined by the equations $\tilde{\varphi}_i(x) = 0$, $\tilde{\varphi} = \langle a^i, x \rangle + \tau_i(x)$, $x \in \mathbb{R}^2$, $\tau_i(x) = o(\|x\|)$ and the constant vectors a^i are the same as described in $(A1)$.

Without loss of generality, we may assume that $\gamma_i \neq \frac{\pi}{2}j$, $j = 1, 3$. Using the transformation $x_1 = r \cos \phi$, $x_2 = r \sin \phi$, equation of the curve c_i can be written, for sufficiently small r, as follows [6]

$$c_i : \phi = \gamma_i + \psi_i(r, \phi), \ i = \overline{1, p}, \tag{3.2.25}$$

where ψ_i is a 2π-periodic function in ϕ, continuously differentiable and $\psi_i = O(r)$. Then, we can define the region situated between the surfaces $\mathscr{S}_{i-1}$ and $\mathscr{S}_i$ as follows:

$$\tilde{D}_1 = \{(r, \phi, z) \mid r \geq 0, \ \gamma_p + \psi_p(r, \phi) < \phi \leq \gamma_1 + 2\pi + \psi_1(r, \phi), \ z \in \mathbb{R}\},$$

$$\tilde{D}_i = \{(r, \phi, z) \mid r \geq 0, \ \gamma_{i-1} + \psi_{i-1}(r, \phi) < \phi \leq \gamma_i + \psi_i(r, \phi), \ z \in \mathbb{R}\}, \text{ where}$$
$i = \overline{2, p}$.

Let ε be a positive number and $N_\varepsilon(\tilde{D}_i)$ denote the ε-neighborhoods of the regions $\tilde{D}_i$, $i = \overline{1, p}$. In addition to $(P1)$, we assume the following list of conditions.

(P2)   Let f_i, h_i, $i = \overline{1, p}$, be functions defined on $N_\varepsilon(\tilde{D}_i)$ and satisfy f_i, $h_i \in C^{(2)}(N_\varepsilon(\tilde{D}_i))$;

(P3) $\tau_i \in C^{(2)}(N_\varepsilon(\tilde{D}_i))$, $i = \overline{1, p}$;

(P4) $f_i(x, z) = o(\|x, z\|)$, $h_i(x, z) = o(\|x, z\|)$, and $f_i(0, z) = 0$, $h_i(0, z) = 0$ for all $z \in \mathbb{R}$, $i = \overline{1, p}$.

We define for $(x, z) \in \tilde{D}_i$, two functions by $\tilde{F}(x, z) = A_i x + f_i(x, z)$ and $\tilde{f}(x, z) = b_i z + h_i(x, z)$, where the matrix A_i and the constant b_i are as defined in $(A2)$ above. In the neighborhood Ω, we consider the following system:

$$\begin{aligned} \frac{dx}{dt} &= \tilde{F}(x, z), \\ \frac{dz}{dt} &= \tilde{f}(x, z). \end{aligned} \tag{3.2.26}$$

Here, it can be easily seen that the functions $\tilde{F}(x, z)$ and $\tilde{f}(x, z)$ have discontinuities on the surfaces $\mathscr{S}_i$, $i = \overline{1, p}$.

For sufficiently small neighborhood Ω, it follows from the conditions $(A1)$ and $(P1)$ that the surfaces $\mathscr{S}_i$ intersect each other only at z-axis, and none of them can intersect itself and $\langle \frac{\partial \tilde{\varphi}_i(x)}{\partial x}, \tilde{F}(x, 0) \rangle \neq 0$ for $x \in c_i$, $i = \overline{1, p}$.

If a solution of system (3.2.26) starts at a point, which is sufficiently close to the origin and on the surface $\mathscr{S}_i$ with fixed i, then this solution can be continued either to the surface $\mathscr{S}_{i+1}$ or $\mathscr{S}_{i-1}$ depending on the direction of the time.

We make use of cylindrical coordinates and rewrite the system (3.2.26) in the following equivalent form

$$
\begin{aligned}
\frac{dr}{d\phi} &= \tilde{G}(r, \phi, z), \\
\frac{dz}{d\phi} &= \tilde{g}(r, \phi, z),
\end{aligned}
\tag{3.2.27}
$$

where $\tilde{G}(r, \phi, z) = \lambda_i r + P_i(r, \phi, z)$ and $\tilde{g}(r, \phi, z) = k_i z + Q_i(r, \phi, z)$ whenever $(r, \phi, z) \in \tilde{D}_i$. The functions P_i and Q_i are 2π-periodic in ϕ, continuously differentiable and $P_i = o(\|(r, z)\|)$, $Q_i = o(\|(r, z)\|)$, $i = \overline{1, p}$.

From the construction, we see that system (3.2.27) is a differential equation with discontinuous right-hand side. For our needs, we redefine the functions $\tilde{G}$ and $\tilde{g}$ in the neighborhoods of the planes $\mathscr{P}_i$, which contain the surface $\mathscr{S}_i$. In other words, we construct new functions G_N and g_N which are continuous everywhere except possibly at the points $(r, \phi, z) \in \mathscr{P}_i$. The redefinition will be made exceptionally at the points which lie between $\mathscr{P}_i$ and $\mathscr{S}_i$ and belong to the regions D_i or D_{i+1} for each i. Therefore, this construction is performed with minimal possible changes corresponding to the B-equivalence method [1], which is the main instrument of our investigation.

It is clear from the context that if $i = p$, then $D_{p+1} = D_1$. Using the argument above, we realize the following reconstruction of the domain. We consider the subregions of D_i and D_{i+1}, which are placed between the plane $\mathscr{P}_i$ and the surface $\mathscr{S}_i$. We refer to the subregions $D_i \cap \tilde{D}_{i+1}$ (light-colored closed regions in Fig. 3.5) and $D_{i+1} \cap \tilde{D}_i$ (dark-colored closed regions in Fig. 3.5) for all i. We extend the

Fig. 3.5 Surfaces $\mathscr{S}_i$, $i = \overline{1, p}$, of discontinuities for the perturbed system (3.2.21)

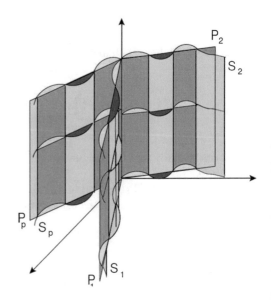

functions $\tilde{G}$ and $\tilde{g}$ from the region $D_i \cap \tilde{D}_{i+1}$ to D_i and from $D_{i+1} \cap \tilde{D}_i$ to D_{i+1} so that the new functions G_N and g_N and their partial derivatives become continuous up to the angle $\phi = \gamma_i$, $i = \overline{1, p}$. According to all these discussions for the definitions of G_N and g_N, we conclude that $G_N(r, \phi, z) = \lambda_i r + P_i(r, \phi, z)$ and $g_N(r, \phi, z) = k_i z + Q_i(r, \phi, z)$ for $(r, \phi, z) \in D_i$. Now, we consider the following differential equation

$$
\begin{aligned}
\frac{dr}{d\phi} &= G_N(r, \phi, z), \\
\frac{dz}{d\phi} &= g_N(r, \phi, z).
\end{aligned}
\tag{3.2.28}
$$

Let us fix $i \in \{1, 2, \ldots, p\}$ and consider a neighborhood of $\mathscr{P}_i$ based on the description above. We shall investigate the following three cases:

I. Assume that the point $(r, \gamma_i, z) \in \tilde{D}_{i+1}$. Let $(r^0(\phi), (z^0(\phi))$ be a solution of (3.2.27) satisfying $(r^0(\gamma_i), (z^0(\gamma_i)) = (\rho, z)$ and ξ_i be the angle where this solution crosses the surface $\mathscr{S}_i$. We denote a solution of (3.2.28) on the interval $[\xi_i, \gamma_i]$ by $(r^1(\phi), z^1(\phi))$ with $(r^1(\xi_i), z^1(\xi_i)) = (r^0(\xi_i), z^0(\xi_i))$. Then

$$
r^0(\phi) = \exp(\lambda_{i+1}(\phi - \gamma_i))\rho + \int_{\gamma_i}^{\phi} \exp(\lambda_{i+1}(\phi - s)) P_{i+1}(r^0(s), s, z^0(s)) ds,
$$

$$
z^0(\phi) = \exp(k_{i+1}(\phi - \gamma_i))z + \int_{\gamma_i}^{\phi} \exp(k_{i+1}(\phi - s)) Q_{i+1}(r^0(s), s, z^0(s)) ds,
$$

and

$$
r^1(\phi) = \exp(\lambda_i(\phi - \xi_i))r^0(\xi_i) + \int_{\xi_i}^{\phi} \exp(\lambda_i(\phi - s)) P_i(r^1(s), s, z^1(s)) ds.
$$

$$
z^1(\phi) = \exp(k_i(\phi - \xi_i))r^0(\xi_i) + \int_{\xi_i}^{\phi} \exp(k_i(\phi - s)) Q_i(r^1(s), s, z^1(s)) ds.
$$

Define a mapping $W_i = (W_i^1, W_i^2)$ on the plane $\phi = \gamma_i$ into itself as follows:

$$
\begin{aligned}
W_i^1(\rho, z) = r^1(\gamma_i) - \rho &= [\exp((\lambda_i - \lambda_{i+1})(\gamma_i - \xi_i)) - 1]\rho \\
&+ \exp(\lambda_i(\gamma_i - \xi_i)) \int_{\gamma_i}^{\xi_i} \exp(\lambda_{i+1}(\xi_i - s)) P_{i+1} ds \\
&+ \int_{\xi_i}^{\gamma_i} \exp(\lambda_i(\gamma_i - s)) P_i ds,
\end{aligned}
$$

$$W_i^2(\rho, z) = z^1(\gamma_i) - z = [\exp((k_i - k_{i+1})(\gamma_i - \xi_i)) - 1]z$$
$$+ \exp(k_i(\gamma_i - \xi_i)) \int_{\gamma_i}^{\xi_i} \exp(k_{i+1}(\xi_i - s))Q_{i+1}ds$$
$$+ \int_{\xi_i}^{\gamma_i} \exp(k_i(\gamma_i - s))Q_i ds.$$

II. If the point $(r, \gamma_i, z) \in \tilde{D}_i$, we can evaluate W_i in the same way:

$$W_i^1(\rho, z) = [\exp((\lambda_i - \lambda_{i+1})(\xi_i - \gamma_i)) - 1]\rho$$
$$+ \exp(\lambda_{i+1}(\gamma_i - \xi_i)) \int_{\gamma_i}^{\xi_i} \exp(\lambda_i(\xi_i - s))P_i ds$$
$$+ \int_{\xi_i}^{\gamma_i} \exp(\lambda_{i+1}(\gamma_i - s))P_{i+1}ds,$$

$$W_i^2(\rho, z) = [\exp((k_i - k_{i+1})(\xi_i - \gamma_i)) - 1]z$$
$$+ \exp(k_{i+1}(\gamma_i - \xi_i)) \int_{\gamma_i}^{\xi_i} \exp(k_i(\xi_i - s))Q_i ds$$
$$+ \int_{\xi_i}^{\gamma_i} \exp(\lambda_{i+1}(\gamma_i - s))Q_{i+1}ds.$$

III. If $(r, \gamma_i, z) \in \mathscr{S}_i$, then $W_i(\rho, z) = 0$.

Results from [6] imply that the functions W_i^1 and W_i^2, $i = \overline{1, p}$, are continuously differentiable and we have $W_i^1 = o(||(\rho, z)||)$, $W_i^2 = o(||(\rho, z)||)$, which follows from the Eq. (3.2.25). In addition, we note that there exists a Lipschitz constant ℓ and a bounded function $m(\ell)$ [1, 6] such that

$$\|W_i^j(\rho_1, z_1) - W_i^j(\rho_2, z_2)\| \le \ell m(\ell)(\|\rho_1 - \rho_2\| + \|z_1 - z_2\|), \quad (3.2.29)$$

for all $\rho_1, \rho_2, z_1, z_2 \in \mathbb{R}$, $j = 1, 2$.

Let $(r(\phi, r_0), z(\phi, z_0))$ be a solution of (3.2.27) with $r(0, r_0) = r_0, z(0, z_0) = z_0$ and ξ_i be the meeting angle of this solution with the surface $\mathscr{S}_i, i = \overline{1, p}$. Denote by $(\xi_i, .\gamma_i]$ the interval $(\xi_i, \gamma_i]$ whenever $\xi_i \le \gamma_i$ and $[\gamma_i, \xi_i)$ if $\gamma_i < \xi_i$. For sufficiently small Ω, the solution $r(\phi, r_0)$, whose trajectory is in Ω for all $\phi \in [0, 2\pi]$, takes the same values with the exception of the oriented intervals $(\xi_i, .\gamma_i]$ as the solution $(\rho(\phi, r_0), z(\phi, z_0))$ with $\rho(0, r_0) = r_0, z(0, z_0) = z_0$ of the impulsive differential equation

$$\frac{d\rho}{d\phi} = G_N(\rho, \phi, z),$$
$$\frac{dz}{d\phi} = g_N(\rho, \phi, z), \qquad \phi \neq \gamma_i,$$
$$\Delta\rho|_{\phi = \gamma_i} = W_i^1(\rho, z),$$
$$\Delta z|_{\phi = \gamma_i} = W_i^2(\rho, z).$$
(3.2.30)

That is, systems (3.2.27) and (3.2.30) are said to be B-equivalent in the sense of the definition in [6]. From the discussion and the construction above, it implies that solutions of (3.2.27) exist in the neighborhood Ω and they are continuous and have discontinuities in the derivative on the surface $\mathscr{S}_i$ for each i. Accordingly, a solution of system (3.2.26) starting at any initial point is continuous, continuously differentiable except possibly at the moments when the trajectories intersect the surface $\mathscr{S}_i$ and is unique.

3.2.4 Center Manifold

We establish a center manifold theorem for sufficiently small solutions to (3.2.30); that is, we show that these solutions can be captured on a two-dimensional invariant manifold and we explicitly describe the dynamics on this manifold.

The functions G_N and g_N in (3.2.30) have been defined as $G_N(r, \phi, z) = \lambda_i r + P_i(r, \phi, z)$ and $g_N(r, \phi, z) = k_i z + Q_i(r, \phi, z)$, where $(r, \phi, z) \in D_i$. Functions P_i and Q_i are 2π-periodic in ϕ and satisfy in a neighborhood of the origin

$$\|P_i(\rho, \phi, z) - P_i(\rho', \phi, z')\| \leq L(\|\rho - \rho'\| + \|z - z'\|), \qquad (3.2.31)$$
$$\|Q_i(\rho, \phi, z) - Q_i(\rho', \phi, z')\| \leq L(\|\rho - \rho'\| + \|z - z'\|), \qquad (3.2.32)$$

for sufficiently small positive constant L, $i = \overline{1, p}$. Applying the methods of the paper [8], we can conclude that system (3.2.30) has two integral manifolds whose equations are given by:

$$\Phi_0(\phi, \rho) = \int_{-\infty}^{\phi} e^{k(\phi - s)} Q(\rho(s, \phi, \rho), s, z(s, \phi, \rho)) ds$$
$$+ \sum_{\gamma_i < \phi} e^{k_i(\phi - \gamma_i)} W_i^2(\rho(\gamma_i, \phi, \rho), z(\gamma_i, \phi, \rho)), \qquad (3.2.33)$$

and

$$\Phi_-(\phi, z) = -\int_{\phi}^{\infty} e^{\lambda(\phi - s)} P(\rho(s, \phi, z), s, z(s, \phi, z)) ds$$
$$+ \sum_{\gamma_i < \phi} e^{\lambda_i(\phi - \gamma_i)} W_i^1(\rho(\gamma_i, \phi, z), z(\gamma_i, \phi, z)), \qquad (3.2.34)$$

where $k = k_i, \lambda = \lambda_i, P = P_i$, and $Q = Q_i$ whenever $(s, \cdot, \cdot) \in D_i$. The pair $(\rho(s, \phi, \rho), z(s, \phi, \rho))$ in (3.2.33), denotes a solution of (3.2.30) satisfying $\rho(\phi, \phi, \rho) = \rho$, and $(\rho(s, \phi, z), z(s, \phi, z))$, in (3.2.34), is a solution of (3.2.30) with $z(\phi, \phi, z) = z$.

It is also shown in [8] that there exist positive constants K_0, M_0, σ_0 such that Φ_0 satisfies:

$$\Phi_0(\phi, 0) = 0, \tag{3.2.35}$$

$$\|\Phi_0(\phi, \rho_1) - \Phi_0(\phi, \rho_2)\| \leq K_0 \ell \|\rho_1 - \rho_2\|, \tag{3.2.36}$$

for all ρ_1, ρ_2 such that a solution $w(\phi) = (\rho(\phi), z(\phi))$ of (3.2.30) with $w(\phi_0) = (\rho_0, \Phi_0(\phi_0, \rho_0)), \rho_0 \geq 0$, is defined on $\mathbb{R}$ and has the following property:

$$\|w(\phi)\| \leq M_0 \rho_0 e^{-\sigma_0(\phi - \phi_0)}, \quad \phi \geq \phi_0. \tag{3.2.37}$$

Furthermore, it is shown that there exist positive constants K_-, M_-, σ_- such that Φ_- satisfies:

$$\Phi_-(\phi, 0) = 0, \tag{3.2.38}$$

$$\|\Phi_-(\phi, z_1) - \Phi_-(\phi, z_2)\| \leq K_- \ell \|z_1 - z_2\|, \tag{3.2.39}$$

for all z_1, z_2 such that a solution $w(\phi) = (\rho(\phi), z(\phi))$ of (3.2.30) with $w(\phi_0) = (\Phi_-(\phi_0, z_0), z_0), z_0 \in \mathbb{R}$, is defined on $\mathbb{R}$ and satisfies

$$\|w(\phi)\| \leq M_- \|z_0\| e^{-\sigma_-(\phi - \phi_0)}, \quad \phi \leq \phi_0. \tag{3.2.40}$$

Denote $S_0 = \{(\rho, \phi, z) : z = \Phi_0(\phi, \rho)\}$ and $S_- = \{(\rho, \phi, z) : \rho = \Phi_-(\phi, z)\}$. Here, S_0 is said to be the *center manifold* and S_- is said to be the *stable manifold*.

The following lemmas can be proven in a similar manner to the ones in [8] with slight changes.

Lemma 3.2.4 *If the Lipschitz constant ℓ is sufficiently small, then for every solution $w(\phi) = (\rho(\phi), z(\phi))$ of (3.2.30) there exists a solution $\mu(\phi) = (u(\phi), v(\phi))$ on the center manifold, S_0, such that*

$$\begin{aligned} \|\rho(\phi) - u(\phi)\| &\leq 2M_0 \|\rho(\phi_0) - u(\phi_0)\| e^{-\sigma_0(\phi - \phi_0)}, \\ \|z(\phi) - v(\phi)\| &\leq M_0 \|z(\phi_0) - v(\phi_0)\| e^{-\sigma_0(\phi - \phi_0)}, \quad \phi \geq \phi_0, \end{aligned} \tag{3.2.41}$$

where M_0 and σ_0 are the constants used in (3.2.37).

Lemma 3.2.5 *For sufficiently small Lipschitz constant ℓ, the surface S_0 is stable in large.*

The dynamics reduced to local, the center manifold S_0 is governed by an impulsive differential equation that is satisfied by the first coordinate of the solutions of (3.2.30) and has the form:

$$\frac{d\rho}{d\phi} = G_N(\rho, \phi, \Phi_0(\phi, \rho)), \quad \phi \neq \gamma_i,$$

$$\Delta\rho|_{\phi=\gamma_i} = W_i^1(\rho, \Phi_0(\phi, \rho)). \tag{3.2.42}$$

The following theorem follows from the reduction principle.

Theorem 3.2.1 *Assume that the conditions assumed so far are fulfilled. Then the trivial solution of (3.2.30) is stable, asymptotically stable or unstable if the trivial solution of (3.2.42) is stable, asymptotically stable or unstable, respectively.*

Using B-equivalence, one can see that the following theorem holds:

Theorem 3.2.2 *Assume that the conditions given above are fulfilled. Then the trivial solution of (3.2.26) is stable, asymptotically stable or unstable if the trivial solution of (3.2.42) is stable, asymptotically stable or unstable, respectively.*

3.2.5 Bifurcation of Periodic Solutions

The center manifold reduction in the previous section allows us to establish a Hopf bifurcation theorem, yielding a very powerful tool to perform a bifurcation analysis on parameter-dependent versions of the considered systems. During the last two decades, many authors have contributed toward developing the general theory.

In order to state the Hopf bifurcation theorem, we include parameter dependence into our framework. In particular, the bifurcation of periodic solutions under the influence of a single parameter μ, $\mu \in (-\mu_0, \mu_0)$, μ_0 a positive constant, is considered for the system:

$$\frac{dx}{dt} = \hat{F}(x, z, \mu),$$
$$\frac{dz}{dt} = \hat{f}(x, z, \mu), \tag{3.2.43}$$

where $\hat{F}(x, z, \mu) = A_i x + f_i(x, z) + \mu F_i(x, z, \mu)$ and $\hat{f}(x, z, \mu) = b_i z + h_i(x, z) + \mu H_i(x, z, \mu)$ whenever $(x, z) \in \tilde{D}_i(\mu) \subset \mathbb{R}^3$, which will be defined below. We will need the following assumptions on the system (3.2.43):

(H1) Let $\{\mathscr{S}_i(\mu)\}_{i=1}^p$ be a collection of surfaces in Ω which start at the z-axis, i.e., $\mathscr{S}_i(\mu) = c_i(\mu) \times \mathbb{R}$, where $c_i(\mu)$ are curves given by $\langle a^i, x \rangle + \tau_i(x) + \mu\kappa_i(x, \mu) = 0$, $x \in \mathbb{R}^2$, $i = \overline{1, p}$;

(H2) Let $\{\mathscr{P}_i(\mu)\}_{i=1}^p$ be a union of half-planes which start at the z-axis, i.e., $\mathscr{P}_i(\mu) = l_i(\mu) \times \mathbb{R}$, where $l_i(\mu)$ is defined by $\langle a^i + \mu\frac{\partial \kappa_i(0, \mu)}{\partial x}, x \rangle = 0$, $i = \overline{1, p}$. Denote by $\gamma_i(\mu)$ the angle of the line $l_i(\mu)$, $i = 1, 2, \ldots, p$.

Like the construction of the regions D_i and $\tilde{D}_i$, we define for $\mu \in (-\mu_0, \mu_0)$, $i = 2, 3, \ldots, p$, the ones associated with the system (3.2.43):

$$\tilde{D}_1(\mu) = \{(r, \phi, z, \mu) \mid r \geq 0, \ \gamma_p(\mu) + \Psi_p < \phi \leq \gamma_1(\mu) + 2\pi + \Psi_1, \ z \in \mathbb{R}\},$$

$$\tilde{D}_i(\mu) = \{(r, \phi, z, \mu) \mid r \geq 0, \ \gamma_{i-1}(\mu) + \Psi_{i-1} < \phi \leq \gamma_i(\mu) + \Psi_i, \ z \in \mathbb{R}\},$$

$$D_1(\mu) = \{(r, \phi, z, \mu) \mid r \geq 0, \ \gamma_p(\mu) < \phi \leq \gamma_1(\mu) + 2\pi, \ z \in \mathbb{R}\},$$

$$D_i(\mu) = \{(r, \phi, z, \mu) \mid r \geq 0, \ \gamma_{i-1}(\mu) < \phi \leq \gamma_i(\mu), \ z \in \mathbb{R}\}.$$

Here, the functions $\Psi_i = \Psi_i(r, \phi, \mu)$ are 2π-periodic in ϕ, continuously differentiable, $\Psi_i = O(r)$, $i = \overline{1, p}$, and can defined in a similar manner to ψ_i in (3.2.25).

To establish the Hopf bifurcation theorem, we also need the following assumptions:

(H3)   The functions $F_i \colon N_\varepsilon(\tilde{D}_i(\mu)) \to \mathbb{R}^2$ and κ_i are analytical functions in x, z, and μ in the ε-neighborhood of their domains;

(H4) $F_i(0, 0, \mu) = 0$ and $\kappa_i(0, \mu) = 0$ hold uniformly for $\mu \in (-\mu_0, \mu_0)$;

(H5)   The matrices A_i, the constants b_i, the functions f_i g_i, τ_i, and the constant vectors a^i correspond to the ones described in systems (3.2.21) and (3.2.26).

In cylindrical coordinates, system (3.2.43) reduces to

$$\frac{dr}{d\phi} = \hat{G}(r, \phi, z, \mu),$$
$$\frac{dz}{d\phi} = \hat{g}(r, \phi, z, \mu),$$
(3.2.44)

$\hat{G}(r, \phi, z, \mu) = \lambda_i(\mu)r + P_i(r, \phi, z, \mu)$ and $\hat{g}(r, \phi, z, \mu) = k_i(\mu)z + Q_i$
(r, ϕ, z, μ) if $(r, \phi, z, \mu) \in \tilde{D}_i(\mu)$.

Let the following impulse system

$$\frac{d\rho}{d\phi} = \hat{G}_N(\rho, \phi, z, \mu),$$
$$\frac{dz}{d\phi} = \hat{g}_N(\rho, \phi, z, \mu), \qquad \phi \neq \gamma_i(\mu),$$
$$\Delta\rho|_\phi = \gamma_i(\mu) = W_i^1(\rho, z, \mu)$$
$$\Delta z|_\phi = \gamma_i(\mu) = W_i^2(\rho, z, \mu)$$
(3.2.45)

be B-equivalent to (3.2.44), where $\hat{G}_N$ and $\hat{g}_N$ stand, respectively, for the extensions of $\hat{G}$ and $\hat{g}$. That is, $\hat{G}_N(\rho, \phi, z, \mu) = \lambda_i(\mu)\rho + P_i(\rho, \phi, z, \mu)$ and $\hat{g}_N(\rho, \phi, z, \mu) = k_i(\mu)z + Q_i(\rho, \phi, z, \mu)$ for $(\rho, \phi, z, \mu) \in D_i(\mu)$. Then, the functions $\hat{G}_N$ and $\hat{g}_N$ and their partial derivatives become continuous up to the angle $\phi = \gamma_i(\mu)$ for $i = \overline{1, p}$. The functions $W_i^1(\rho, z, \mu)$ and $W_i^2(\rho, z, \mu)$ can be defined in the same manner as in Sect. 3.1.

Following the same methods which are used to obtain (3.2.33) and (3.2.34), we can say that system (3.2.45) has two integral manifolds whose equations are given by:

$$\Phi_0(\phi, \rho, \mu) = \int_{-\infty}^{\phi} e^{k(\mu)(\phi-s)} Q(\rho(s, \phi, \rho, \mu), s, z(s, \phi, \rho, \mu), \mu) ds$$

$$+ \sum_{\gamma_i(\mu)<\phi} e^{k_i(\mu)(\phi-\gamma_i(\mu))} W_i^2(\rho(\gamma_i(\mu), \phi, \rho, \mu), z(\gamma_i(\mu), \phi, \rho, \mu), \mu), \quad (3.2.46)$$

and

$$\Phi_-(\phi, z, \mu) = -\int_{\phi}^{\infty} e^{\lambda(\mu)(\phi-s)} P(\rho(s, \phi, z, \mu), s, z(s, \phi, z, \mu), \mu) ds$$

$$+ \sum_{\gamma_i(\mu)<\phi} e^{\lambda_i(\mu)(\phi-\gamma_i(\mu))} W_i^1(\rho(\gamma_i(\mu), \phi, z, \mu), z(\gamma_i(\mu), \phi, z, \mu), \mu),$$

$$(3.2.47)$$

where $k(\mu) = k_i(\mu)$, $\lambda(\mu) = \lambda_i(\mu)$, $P = P_i$, and $Q = Q_i$ whenever $(s, \cdot, \cdot, \cdot) \in D_i(\mu)$. In (3.2.46), the pair $(\rho(s, \phi, \rho, \mu), z(s, \phi, \rho, \mu))$ denotes a solution of (3.2.45) satisfying $\rho(\phi, \phi, \rho, \mu) = \rho$. Similarly, $(\rho(s, \phi, z, \mu), z(s, \phi, z, \mu))$, in (3.2.47), is a solution of (3.2.45) with $z(\phi, \phi, z, \mu) = z$.

Set $S_0(\mu) = \{(\rho, \phi, z, \mu) : z = \Phi_0(\phi, \rho, \mu)\}$ and $S_-(\mu) = \{(\rho, \phi, z, \mu) : \rho = \Phi_-(\phi, z, \mu)\}$.

The reduced system on the center manifold $S_0(\mu)$ is given by

$$\frac{d\rho}{d\phi} = \hat{G}_N(\rho, \phi, \Phi_0(\phi, \rho, \mu), \mu), \quad \phi \neq \gamma_i(\mu),$$
$$\Delta\rho \mid_{\phi=\phi_i(\mu)} = W_i^1(\rho, \Phi_0(\phi, \rho, \mu), \mu). \qquad (3.2.48)$$

Similar to (3.2.23) and (3.2.24), we can define the functions

$$q_1(\mu) = \exp\left(\sum_{i=1}^{p} \lambda_i(\mu)\theta_i(\mu)\right), \qquad (3.2.49)$$

$$q_2(\mu) = \exp\left(\sum_{i=1}^{p} k_i(\mu)\theta_i(\mu)\right). \qquad (3.2.50)$$

System (3.2.48) is a system of the type studied in [6], and there it is shown that this system, for sufficiently small μ, has a periodic solution with period 2π. For our needs, we shall show that if the first coordinate of a solution of (3.2.45) is 2π-periodic, then so is the second one.

Now, since

$$\rho(s + 2\pi, \phi + 2\pi, \rho, \mu) = \rho(s, \phi, \rho, \mu),$$

$$z(s + 2\pi, \phi + 2\pi, \rho, \mu) = z(s, \phi, \rho, \mu),$$

and each Q_i is 2π-periodic in ϕ, we have

$$
\begin{aligned}
&\Phi_0(\phi + 2\pi, \rho, \mu) \\
&= \int_{-\infty}^{\phi+2\pi} e^{k(\mu)(\phi+2\pi-s)} Q(\rho(s, \phi+2\pi, \rho, \mu), s, z(s, \phi+2\pi, \rho, \mu), \mu) ds \\
&\quad + \sum_{\gamma_i(\mu)<\phi+2\pi} e^{k_i(\mu)(\phi+2\pi-\gamma_i(\mu))} \times \\
&\qquad\qquad \times W_i^2(\rho(\gamma_i(\mu), \phi+2\pi, \rho, \mu), z(\gamma_i(\mu), \phi+2\pi, \rho, \mu), \mu) \\
&= \int_{-\infty}^{\phi} e^{k(\mu)(\phi-t)} Q(\rho(t, \phi, \rho, \mu), t, z(t, \phi, \rho, \mu), \mu) dt \\
&\quad + \sum_{\bar{\gamma}_i(\mu)<\phi} e^{k_i(\mu)(\phi-\bar{\gamma}_i(\mu))} W_i^2(\rho(\bar{\gamma}_i(\mu), \phi, \rho, \mu), z(\bar{\gamma}_i(\mu), \phi, \rho, \mu), \mu) \\
&= \Phi_0(\phi, \rho, \mu),
\end{aligned}
$$

where the substitutions $s = t + 2\pi$ and $\gamma_i(\mu) = \bar{\gamma}_i(\mu) + 2\pi$ are used for the integral and summation in the second equality.

Then, we obtain the following theorem whose proof for two-dimensional case can be found in [6].

Theorem 3.2.3 *Assume that $q_1(0) = 1, q_1'(0) \neq 0, q_2(0) < 1$, and the origin is a focus for (3.2.26). Then, for sufficiently small r_0 and z_0, there exists a unique continuous function $\mu = \delta(r_0, z_0), \delta(0, 0) = 0$ such that the solution $(r(\phi, \delta(r_0, z_0)), z(\phi, \delta(r_0, z_0)))$ of (3.2.44), with the initial condition $(r(0, \delta(r_0, z_0), z(0, \delta(r_0, z_0)) = (r_0, z_0)$, is periodic with period 2π. The period of the corresponding periodic solution of (3.2.43) is $\sum_{i=1}^{p} \frac{\theta_i}{\beta_i} + o(|\mu|)$.*

3.2.6 An Example

For convenience, in this section, we shall use the corresponding notations that are adopted above.

Example 3.2.1 Let $c_1(\mu)$ and $c_2(\mu)$ denote the curves determined by $x_2 = \frac{1}{\sqrt{3}}x_1 + (1 + \mu)x_1^3, x_1 > 0$ and $x_2 = \sqrt{3}x_1 + x_1^5 + \mu x_1^2, x_1 < 0$, respectively. We choose

$$
A_1 = \begin{bmatrix} -0.7 & -2 \\ 2 & -0.7 \end{bmatrix}, \quad f_1(x, z) = \begin{bmatrix} x_1 z \sqrt{x_1^2 + x_2^2} \\ x_2 z^2 \sqrt{x_1^2 + x_2^2} \end{bmatrix},
$$

$$
F_1(x, z, \mu) = \begin{bmatrix} x_1(1 + z) \\ x_2 \end{bmatrix},
$$

$$b_1 = 2, \ \ h_1(x, z) = x_1^2 z, \ \ H_1(x, z, \mu) = z,$$

$$A_2 = \begin{bmatrix} 0.5 & -2 \\ 2 & 0.5 \end{bmatrix}, \ \ f_2(x, z) = \begin{bmatrix} -2x_1 z^2 \sqrt{x_1^2 + x_2^2} \\ -2x_2 \sqrt{x_1^2 + x_2^2} \end{bmatrix},$$

$$F_2(x, z, \mu) = \begin{bmatrix} x_1 \\ x_2(1 + x_1 z) \end{bmatrix},$$

$$b_2 = -1.5, \ \ h_2(x, z) = x_1 z, \ \ H_2(x, z, \mu) = [1 - (x_1^2 + x_2^2)]z.$$

After these preparations, we consider the system

$$\frac{dx}{dt} = \hat{F}(x, z, \mu),$$
$$\frac{dz}{dt} = \hat{f}(x, z, \mu),$$
(3.2.51)

where $\quad \hat{F}(x, z, \mu) = A_i x + f_i(x, z) + \mu F_i(x, z, \mu) \quad$ and $\quad \hat{f}(x, z, \mu) = b_i z + h_i(x, z) + \mu H_i(x, z, \mu)$ whenever $(x, z) \in \tilde{D}_i(\mu)$.

Since $l_1(\mu)$ $(l_2(\mu))$ coincides with l_1 (l_2), $\gamma_1 = \gamma_1(\mu) = \frac{\pi}{6}$ and $\gamma_2 = \gamma_2(\mu) = \frac{4\pi}{3}$.
Now, we can evaluate $q_1(\mu)$ and $q_2(\mu)$ as follows:

$$q_1(\mu) = \exp(\pi\mu), \tag{3.2.52}$$

$$q_2(\mu) = \exp(\pi(\mu - \frac{1}{24})). \tag{3.2.53}$$

From (3.2.52) and (3.2.53), we can see that $q_1(0) = 1$, $q_1'(0) > 0$ and $q_2(0) < 1$. Therefore, by Theorem 3.2.3, system (3.2.51) has a periodic solution with period $\approx \pi$. One can see from the Figs. 3.6 and 3.7, which are obtained for different initial conditions, that the trajectories approach to the periodic solution from above and below. In other words, system (3.2.51) admits a limit cycle.

Fig. 3.6 Existence of a periodic solution for (3.2.51)

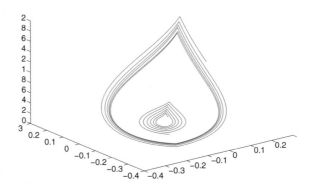

Fig. 3.7 Existence of another periodic solution for (3.2.51) with a different initial value

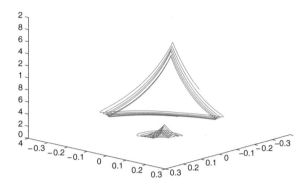

3.3 Notes

Differential equations whose right-hand sides are discontinuous on nonlinear surfaces were investigated in [46, 54, 55] by the method of B-equivalence. This method was first proposed to reduce impulsive systems with variable time of impulses to the systems with fixed moments of impulse effects. It then turned out that this method could be used for differential equations with discontinuous right-hand side as well [46, 54]. That is, through the B-equivalence method, differential equations with discontinuous vector fields with nonlinear discontinuity sets can be reduced to impulsive differential equations with fixed moments of impulses.

Dankowicz and Nordmark [102] study bifurcations of stick-slip oscillations for the friction model which leads to a nonsmooth dynamical system having discontinuity at the first derivative of the vector field. Feigin [112, 113] considers C-bifurcations, also known as border-collision bifurcations, in Filippov systems being a subclass of discontinuous systems described by differential equations with a discontinuous right-hand side [115]. Border-collision bifurcations for nonsmooth discrete maps are also addressed by Nusse and Yorke [183, 184].

Hopf bifurcation for smooth systems is characterized by a pair of complex conjugate eigenvalues of the linearized system. It is well known that it is not the case for systems of differential equations with discontinuities. Although the system specified in (3.1.14) together with the assumption ($A2$) reflects a special class of such systems, it is worthwhile to develop a technique for the investigation of bifurcation problem as it exhibits complicated bifurcation phenomena. Further, the problem can be generalized by taking the matrices A_i, $i = 1, 2, \ldots, p$, not only of focus type in all subregions but also of another types; e.g., they may be hyperbolic with real eigenvalues. Clearly, this problem can be analyzed in a similar way when it is required by concrete applications in mechanics, electronics, biology, etc. The present chapter contains mainly the results of papers [17, 20].

Chapter 4
Nonautonomous Bifurcation in Impulsive Bernoulli Equations

4.1 The Transcritical and the Pitchfork Bifurcations

In this chapter, we discuss impulsive generalizations of the nonautonomous pitchfork and transcritical bifurcations. Scalar differential equations with fixed moments of impulses are considered. By means of certain systems, we show that how the idea of pullback attracting sets remains a fruitful concept in the impulsive systems. Basics of the theory are provided.

4.1.1 Introduction

Asymptotic behavior of a solution near a fixed point and analysis of bifurcation is of a great importance in the qualitative theory of differential equations. In autonomous ordinary differential equations, this theory is well developed. As in the autonomous systems, nonautonomous bifurcation is understood as a qualitative change in the structure and stability of the invariant sets of the system. However, to implement this concept in nonautonomous systems, locally defined notions of attractive and repulsive solutions are needed. There are currently qualitative studies which are devoted to nonautonomous bifurcation theory by treating pullback attractors [75, 76, 138, 139, 142, 152, 154, 204]. In the classical theory of stability, one is interested in the asymptotic behavior of a solution as $t \to \infty$ for a fixed t_0, which is called forward attraction. On the other hand, the theory of pullback attraction deals with the asymptotic behavior of the solution as $t_0 \to -\infty$ for a fixed t [61, 75, 81, 85, 96–98, 142, 151, 153, 155, 227]. These two types of attractions give the same convergence analysis for autonomous dynamical systems. The approach of pullback attraction is required for the discussion of bifurcation analysis in nonautonomous differential equations by defining corresponding types of stability.

Modeling problems in the states of dynamical systems with time-dependent vector fields leads to nonautonomous problems. Moreover, these models may depend on

© Springer Nature Singapore Pte Ltd. and Higher Education Press 2017
M. Akhmet and A. Kashkynbayev, *Bifurcation in Autonomous and Nonautonomous Differential Equations with Discontinuities*,
Nonlinear Physical Science, DOI 10.1007/978-981-10-3180-9_4

some parameters which are accepted as the influence of an environment. In this case, it is an interesting issue to analyze qualitative changes when these parameters are varied. The main object of nonautonomous bifurcation theory is concerned in describing these changes. In addition to these, there may be abrupt changes at prescribed times in the real-world evolutionary processes. These progressions are portrayed as impulsive phenomena [1, 103, 156, 156, 216], which are in no way, shape or form, however, regular in modeling in mechanics, electronics, biology, neural networks, medicine, and in social sciences [1, 3, 19, 43]. Hence, an impulsive differential equation is recognized as one of the central apparatuses to better comprehend the function of discontinuity in this present reality issues. Extending nonautonomous bifurcation theory to impulsive systems is a contemporary problem.

4.1.2 Preliminaries

In this section, we introduce concepts of attractive and repulsive solutions, which are used to analyze asymptotic behavior of impulsive nonautonomous systems. This chapter is concerned with systems of the type

$$\begin{aligned} x' &= f(t, x), \\ \Delta x|_{t=\theta_i} &= J_i(x), \end{aligned} \tag{4.1.1}$$

where $\Delta x|_{t=\theta_i} := x(\theta_i+) - x(\theta_i)$, $x(\theta_i+) = \lim_{t \to \theta_i^+} x(t)$. The system (4.1.1) is defined on the set $\Omega = \mathbb{R} \times \mathbb{Z} \times G$ where $G \subseteq \mathbb{R}^n$. θ is a nonempty sequence with the set of indexes $\mathbb{Z}$, set of integers, such that $|\theta_i| \to \infty$ as $|i| \to \infty$. Let $\phi(t, t_0, x_0)$ be solution of (4.1.1). In this chapter, we deal with scalar impulsive differential equations such that $\phi(t, t_0, x_0)$ is noncontinuable. Solutions are unique both forward and backward in time.

We say that the function $\phi : \mathbb{R} \to \mathbb{R}^n$ is from the set $PC(\mathbb{R}, \theta)$, where $\theta = \{\theta_i\}$ is an infinite set such that $|\theta_i| \to \infty$ as $|i| \to \infty$, if:

- ϕ is left continuous on $\mathbb{R}$;
- it is continuous everywhere except possibly points of θ where it has discontinuities of the first kind.

One cannot follow the same way in developing the theory for impulsive differential equations as for autonomous systems because there are certain problems. Namely, there may not be any equilibrium point at all. That is, it is hard to find a point x_0 which satisfies both $f(x_0, t) = 0$ for all $t \in \mathbb{R}$ and $J_i(x_0) = 0$ for all $i \in \mathbb{Z}$. Therefore, the notion of equilibrium point is replaced with a bounded solution or a complete trajectory, which is a particular example of invariant sets. We investigate appearances and disappearances of complete trajectories that are stable and unstable in the pullback sense.

A time-varying family of set $\mathfrak{A}(t)$ is invariant if $x_0 \in \mathfrak{A}(t_0)$ implies that $\phi(t, t_0, x_0) \in \mathfrak{A}(t)$. In order to study nonautonomous bifurcation with impulses, we should define corresponding concepts of stability. In this chapter, we use Hausdorff semi-distance between sets X and Y as $d(X, Y) = \sup_{x \in X} \inf_{y \in Y} d(x, y)$

4.1.2.1 Attraction and Stability

Asymptotic properties of continuous dynamics and dynamics with discontinuity are the same. Therefore, we shall use notion of pullback attracting sets without any change from [61, 75, 81, 85, 96–98, 140, 142, 151, 153, 155, 204, 227] and references therein. In autonomous system, to ensure that an invariant set $\mathfrak{A}$ is attracting it is enough to have the existence of a neighborhood N of $\mathfrak{A}$ such that $d(\phi(t, 0, x_0), \mathfrak{A}) \to 0$ as $t \to \infty$. In autonomous system, asymptotic behavior of dynamics relies on given $t - t_0$ rather than the initial time only. Hence, the idea of attraction for autonomous systems is identical to the presence of a neighborhood N of $\mathfrak{A}$ for each fixed $t \in \mathbb{R}$, $d(\phi(t, t_0, x_0), \mathfrak{A}) \to 0$ as $t_0 \to -\infty$ for all $x_0 \in N$. This is the principle thought under the pullback attraction [140, 227]. That is, we are interested in asymptotic behavior as $t_0 \to -\infty$ for fixed t, which makes it possible to analyze time-dependent sets.

Definition 4.1.1 ([140]) An invariant set $\mathfrak{A}(t)$ is called pullback attracting if for every $t \in \mathbb{R} \lim_{t_0 \to -\infty} d(\phi(t, t_0, x_0), \mathfrak{A}(t)) = 0$.

Having given meanings of pullback attraction, one needs to characterize related ideas of stability, instability, and asymptotic stability in order to investigate asymptotic analysis in the pullback sense. Next, we start with defining stability in the pullback sense.

Definition 4.1.2 ([152]) An invariant set $\mathfrak{A}(t)$ is pullback stable if for every $t \in \mathbb{R}$ and $\varepsilon > 0$ there exists a $\delta(t) > 0$ such that for any $t_0 < t, x_0 \in N(\mathfrak{A}(t_0), \delta(t))$ implies that $\phi(t, t_0, x_0) \in N(\mathfrak{A}(t), \varepsilon)$.

An invariant set $\mathfrak{A}(t)$ is said to be pullback asymptotically stable if it is pullback stable and pullback attracting. As we are busy with scalar impulsive systems, one can verify that pullback attraction implies pullback stability for bounded trajectories. Next, we state the following lemma which will be useful in what follows.

Lemma 4.1.1 *Let $y(t)$ be a locally pullback attracting complete trajectory of a scalar impulsive system. Then, $y(t)$ is also pullback stable.*

The proof of this lemma, given by Langa et al. in [154], for continuous case is the same for impulsive systems. Thus, the last lemma allows us to concentrate on only pullback attraction properties of a complete trajectory instead of carrying out pullback stability.

As one would expect, pullback instability is characterized through the converse of pullback stability. That is, an invariant set $\mathfrak{A}(t)$ is called pullback unstable if it is not pullback stable, i.e., if there exists a $t \in \mathbb{R}$ and $\varepsilon > 0$ such that for each $\delta > 0$, there exists a $t_0 < t$ and $x_0 \in N(\mathfrak{A}(t_0), \delta)$ such that $d(\phi(t, t_0, x_0), \mathfrak{A}(t)) > \varepsilon$. However, the notion of unstable set, which Crauel defined for the random dynamical systems, seems to be more natural instrument in discontinuous dynamics point of view.

Definition 4.1.3 ([96]) The unstable set, $U_{\mathfrak{A}(t)}$, of an invariant set $\mathfrak{A}(t)$ is defined as

$$U_{\mathfrak{A}(t)} = \{u : \lim_{t \to -\infty} d(\phi(t, t_0, u), \mathfrak{A}(t)) = 0\}.$$

We say that $\mathfrak{A}(t)$ is asymptotically unstable if the relation $U_{\mathfrak{A}(t)} \neq \mathfrak{A}(t)$ is fulfilled for some t.

If $\mathfrak{A}(t)$ is invariant then one can see that $\mathfrak{A}(t) \subset U_{\mathfrak{A}(t)}$ is satisfied. Thus, from the last definition we have that $\mathfrak{A}(t)$ is strict subset of $U_{\mathfrak{A}(t)}$. In the sequel, we need the following result.

Proposition 4.1.1 ([152]) *If $\mathfrak{A}(t)$ is asymptotically unstable, then it is also locally pullback unstable and cannot be locally pullback attracting.*

This result proven by Langa et al. in [152] is valid for impulsive systems. Thus, we omit the proof and refer to [152]. Note that the idea of the asymptotic instability is a definition of time-reversed forward attraction. Alternatively, it is conceivable to define instability as a time-reversed version of pullback attraction.

Definition 4.1.4 ([154]) An invariant set $\mathfrak{A}(t)$ is pullback repelling if it is pullback attracting for time-reversed system, i.e., if for every $t \in \mathbb{R}$ and every $x_0 \in \mathbb{R}^n$,

$$\lim_{t_0 \to \infty} d(\phi(t, t_0, x_0), \mathfrak{A}(t)) = 0.$$

4.1.3 The Pitchfork Bifurcation

In this section, we consider the following system

$$x' = p(t)x - q(t)x^3, \tag{4.1.2a}$$

$$\Delta x|_{t=\theta_i} = -x + \frac{x}{\sqrt{c_i + d_i x^2}}, \tag{4.1.2b}$$

where $p, q \in PC(\mathbb{R}, \theta)$. Assume that there exist constants A, B, C, and D such that

$$|p(t)| < A < \infty \text{ and } 0 < c_i \leq C < \infty, \tag{4.1.3}$$

and

$$0 < b_0 \leq q(t) < B < \infty \text{ and } 0 < d_i \leq D < \infty, \tag{4.1.4}$$

for $i \in \mathbb{Z}$ and $t \in \mathbb{R}$. We suppose that there exist positive numbers $\underline{\theta}$ and $\overline{\theta}$ such that

$$\underline{\theta} \leq \theta_{i+1} - \theta_i \leq \overline{\theta}. \tag{4.1.5}$$

Moreover, there exists the limit

$$\lim_{t-s \to \infty} \frac{2 \int_s^t p(u)du - \sum_{s \le \theta_i < t} \ln c_i}{t - s} = \gamma. \tag{4.1.6}$$

By means of substitution $y = \dfrac{1}{x^2}$, the system (4.1.2) is converted to the impulsive linear nonhomogeneous system

$$\begin{aligned} \dot{y} &= -2p(t)y + 2qt), \\ \Delta y|_{t=\theta_i} &= (c_i - 1)y + d_i. \end{aligned} \tag{4.1.7}$$

In what follows, we discuss the system (4.1.7) to analyze the system (4.1.2). Since $c_i \ne 0$, the transition matrix of the associated homogeneous part of (4.1.7) is, [1, 156, 216],

$$Y(t, s) = e^{-2 \int_s^t p(u)du} \prod_{s \le \theta_i < t} c_i = e^{-\frac{2 \int_s^t p(u)du - \sum_{s \le \theta_i < t} \ln c_i}{t-s}(t-s)}, \ t \ge s. \tag{4.1.8}$$

Lemma 4.1.2 *If $\alpha > \gamma > \beta > 0$, then there exist positive numbers M and m such that*

$$me^{-\alpha(t-s)} \le Y(t, s) \le Me^{-\beta(t-s)}, \ t \ge s. \tag{4.1.9}$$

Proof By relation (4.1.6), there exists T such that if $t - s \ge T$, then

$$\beta < \frac{2 \int_s^t p(u)du - \sum_{s \le \theta_i < t} \ln c_i}{t - s} < \alpha.$$

Consequently, by means of (4.1.3) and (4.1.5), it is true that

$$M = \sup_{0 \le t-s \le T} e^{-2 \int_s^t p(u)du} \prod_{s \le \theta_i < t} c_i$$

and

$$m = \inf_{0 \le t-s \le T} e^{-2 \int_s^t p(u)du} \prod_{s \le \theta_i < t} c_i.$$

Hence,

$$me^{-\alpha(t-s)} \le Y(t, s) = e^{-2 \int_s^T p(u)du + \sum_{s \le \theta_i < T} \ln c_i} e^{-2 \int_T^t p(u)du + \sum_{T \le \theta_i < t} \ln c_i}$$
$$\le Me^{-\beta(t-s)},$$

for $t \ge s$. The lemma is proved. $\square$

Theorem 4.1.1 *Assume that (4.1.3), (4.1.4), and (4.1.6) hold for the system (4.1.2). Then, for $\gamma < 0$ the trivial solution is globally asymptotically pullback stable and for $\gamma > 0$ the trivial solution is asymptotically unstable, and complete trajectories*

$\pm v(t, \gamma)$ are locally asymptotically pullback stable and satisfy the following relation.

$$v^2(t, \gamma) = \frac{1}{2 \int_{-\infty}^{t} Y(t, s)q(s)ds + \sum_{\theta_i < t} Y(t, \theta_i+)d_i}.$$

Proof By substitution $y = \dfrac{1}{x^2}$, we see that the solution of the system (4.1.2) satisfies the following equation, [1, 156, 216],

$$y(t, t_0, y_0) = Y(t, t_0)y_0 + 2 \int_{t_0}^{t} Y(t, s)q(s)ds + \sum_{t_0 \le \theta_i < t} Y(t, \theta_i+)d_i. \quad (4.1.10)$$

By means of (4.1.6), one can see that asymptotic behavior of $y(t, t_0, y_0)$ is to depend on sign of γ.

Consider the case $\gamma < 0$. From (4.1.10), it follows that $y(t, t_0, y_0) \to \infty$ as $t_0 \to -\infty$. Thus, $x(t, t_0, x_0) \to 0$ both as $t_0 \to -\infty$ and as $t \to \infty$. Hence, all solutions are attracted both forward and pullback to the trivial solution.

If $\gamma > 0$, then from (4.1.10) it follows that $y(t, t_0, y_0) \to 0$ as $t \to \infty$ implying that all solutions are unbounded as $t \to \infty$. However, as $t_0 \to -\infty$ we have

$$\lim_{t_0 \to -\infty} y(t, t_0, y_0) = 2 \int_{-\infty}^{t} Y(t, s)q(s)ds + \sum_{\theta_i < t} Y(t, \theta_i+)d_i. \quad (4.1.11)$$

The last equation reply that

$$\lim_{t_0 \to -\infty} x^2(t, t_0, x_0) = v^2(t, \gamma) = \frac{1}{2 \int_{-\infty}^{t} Y(t, s)q(s)ds + \sum_{\theta_i < t} Y(t, \theta_i+)d_i}$$
$$(4.1.12)$$

where $s, \theta_i \in (-\infty, t]$. By means of (4.1.5) and Lemma 4.1.2, one can show that

$$0 < \frac{2mb_0}{\alpha} < 2 \int_{-\infty}^{t} Y(t, s)q(s)ds + \sum_{\theta_i < t} Y(t, \theta_i+)d_i$$
$$< \frac{2BM}{\beta} + DM \sum_{\theta_i < t} e^{-\beta(t-\theta_i)}$$
$$\le \frac{2BM}{\beta} + DM \sum_{i=0}^{\infty} e^{-i\beta\underline{\theta}} \qquad (4.1.13)$$
$$= \frac{2BM}{\beta} + DM \frac{1}{1 - e^{\beta\underline{\theta}}} < \infty.$$

Thus, $v^2(t, \gamma)$ is bounded both from above and from below. To see that $v(t, \gamma)$ is a complete trajectory, it would be enough to check that $\eta(t) = \dfrac{1}{v^2(t, \gamma)}$ satisfies (4.1.7). Indeed,

$$\dot{\eta} = -4p(t)\int_{-\infty}^{t} Y(t,s)q(s)ds + 2Y(t,t)q(t) - 2p(t)\sum_{\theta_i < t} Y(t,\theta_i+)d_i$$

$$= -2p(t)\left\{2\int_{-\infty}^{t} Y(t,s)q(s)ds + \sum_{\theta_i < t} Y(t,\theta_i+)d_i\right\} + 2q(t) \qquad (4.1.14)$$

$$= -2p(t)\eta + 2q(t).$$

To show that $\eta(t)$ satisfies the equation jumps, we note for fixed j it is true that $Y(\theta_j+,s) - Y(\theta_j,s) = (c_j - 1)Y(\theta_j,s)$; so that $Y(\theta_j+,s) = c_j Y(\theta_j,s)$. Then,

$$\Delta\eta(t)|_{t=\theta_j} = \eta(\theta_j+) - \eta(\theta_j)$$

$$= 2\int_{-\infty}^{\theta_j+} Y(\theta_j+,s)q(s)ds + \sum_{\theta_i < \theta_j+} Y(\theta_j+,\theta_j+)d_j$$

$$-2\int_{-\infty}^{\theta_j} Y(\theta_j,s)q(s)ds - \sum_{\theta_i < \theta_j} Y(\theta_j,\theta_j+)d_j$$

$$= 2c_j\int_{-\infty}^{\theta_j} Y(\theta_j,s)q(s)ds - 2\int_{-\infty}^{\theta_j} Y(\theta_j,s)q(s)ds + d_j$$

$$+ \sum_{\theta_i < \theta_j} c_j Y(\theta_j,\theta_j+)d_j - \sum_{\theta_i < \theta_j} Y(\theta_j,\theta_j+)d_j \qquad (4.1.15)$$

$$= (c_j - 1)\left\{2\int_{-\infty}^{\theta_j} Y(\theta_j,s)q(s)ds + \sum_{\theta_i < \theta_j} Y(\theta_j,\theta_j+)d_j\right\} + d_j$$

$$= (c_j - 1)\eta(\theta_j) + d_j.$$

The above analysis shows that $v(t,\gamma)$ is pullback attracting. Thus, Lemma 4.1.1 implies that $v(t,\gamma)$ is pullback stable. By means of (4.1.10), it follows that

$$x^2(t,t_0,x_0) =$$

$$\frac{1}{y(t,t_0,y_0)} = \frac{1}{Y(t,t_0)x_0^{-2} + 2\int_{t_0}^{t} Y(t,s)q(s)ds + \sum_{t_0 \le \theta_i < t} Y(t,\theta_i+)d_i}$$

$$= \frac{1}{Y(t,t_0)(x_0^{-2} - v^{-2}(t_0)) + 2\int_{-\infty}^{t} Y(t,s)q(s)ds + \sum_{\theta_i < t} Y(t,\theta_i+)d_i}.$$

$$(4.1.16)$$

If $|x_0| < v(t_0)$ so that $x^{-2} - v^{-2}(t_0) > 0$, then $x(t)$ converges to 0 as $t \to -\infty$ implying that the origin is asymptotically unstable. This finalizes the proof of the theorem. $\square$

Remark 4.1.1 In the similar manner, it can be easily shown that the results of Theorem 4.1.1 hold for the following system.

$$x' = p(t)x - q(t)x^3,$$
$$\Delta x|_{t=\theta_i} = -x - \frac{x}{\sqrt{c_i + d_i x^2}}.$$

Example 4.1.1 Let $p(t) \equiv a$, $c_i \equiv c$, and $\theta_i = ih$ for the system (4.1.2) with $h > 0$. That is,

$$x' = ax - q(t)x^3,$$
$$\Delta x|_{t=ih} = -x + \frac{x}{\sqrt{c+d_ix^2}}. \tag{4.1.17}$$

Then $\gamma = 2a - \frac{1}{h}\ln c$. By means of $y = \frac{1}{x^2}$, the system (4.1.17) is reduced to the linear impulsive system

$$\dot{y} = -2ay + 2q(t),$$
$$\Delta y|_{t=ih} = (c-1)y + d_i. \tag{4.1.18}$$

Asymptotic behavior of (4.1.18) is to depend on the sign of $2a - \frac{1}{h}\ln c = \gamma$, and results of Theorem 4.1.1 are true for the system (4.1.17). If, in particular, $c = 1$ and $d_i = 0$, then there is no equation of jumps in the system (4.1.17). Moreover, $\gamma = 2a$ so that asymptotic behavior of (4.1.18) is to depend on the sign of a. Thus, results of Theorem 4.1.1 are generalization of the results obtained in the studies of Langa et al. in [152] and Caraballo and Langa in [75].

4.1.4 The Transcritical Bifurcation

Consider the impulsive system

$$x' = p(t)x - q(t)x^2, \tag{4.1.19a}$$
$$\Delta x|_{t=\theta_i} = -x + \frac{x}{c_i + d_ix}, \tag{4.1.19b}$$

where $c_i > 0$, $d_i \in \mathbb{R}$, $i \in \mathbb{Z}$, $p, q \in PC(\mathbb{R}, \theta)$. Differently from the previous section, we do not impose any condition on the function p. However, as in the previous section, we suppose that there exist positive numbers $\underline{\theta}$ and $\overline{\theta}$ such that $\underline{\theta} \leq \theta_{i+1} - \theta_i \leq \overline{\theta}$, and there exists the limit

$$\lim_{t-s\to\infty} \frac{\int_s^t p(u)du - \sum_{s\leq\theta_i<t}\ln c_i}{t-s} = \gamma. \tag{4.1.20}$$

The function q and the numbers d_i are asymptotically positive as $t \to -\infty$ and $\theta_i \to -\infty$, respectively. In other words, there exist constants $\underline{b}$ and $\underline{d}$ such that

$$q(t) \geq \underline{b} > 0 \text{ for all } t \leq T^-, \text{ and } d_i \geq \underline{d} > 0 \text{ for all } \theta_i \leq T^-. \tag{4.1.21}$$

By means of substitution $x = \dfrac{1}{y}$, the system (4.1.19) is reduced to the following impulsive linear nonhomogeneous differential equation.

$$\begin{aligned}\dot{y} &= -p(t)y + q(t),\\ \Delta y|_{t=\theta_i} &= (c_i - 1)y + d_i.\end{aligned} \tag{4.1.22}$$

The transition matrix of the associated homogeneous part of the system (4.1.22) is, [1],

$$Y(t,s) = e^{-\int_s^t p(u)du} \prod_{s \le \theta_i < t} c_i = e^{-\frac{\int_s^t p(u)du - \sum_{s \le \theta_i < t} \ln c_i}{t-s}\,)(t-s)}, \quad t \ge s. \tag{4.1.23}$$

Assume that there exists a $\gamma_0 > 0$ such that

$$0 < m_\gamma \le x_\gamma(t) = \frac{1}{\int_{-\infty}^t Y(t,s)q(s)ds + \sum_{\theta_i < t} Y(t,\theta_i+)d_i} \le M_\gamma \tag{4.1.24}$$

for all $t \in \mathbb{R}$, $i \in \mathbb{Z}$, $0 < \gamma < \gamma_0$, and

$$\liminf_{t_0 \to -\infty} \frac{Y(t,t_0)}{\int_{t_0}^t Y(t,s)q(s)ds + \sum_{t_0 \le \theta_i < t} Y(t,\theta_i+)d_i} \ge m_\gamma > 0 \tag{4.1.25}$$

for all $-\gamma_0 < \gamma < 0$.

Theorem 4.1.2 *Assume that the above conditions hold for Eq. (4.1.19). Then, for $-\gamma_0 < \gamma < 0$ the origin is locally pullback attracting in $\mathbb{R}$ and for $0 < \gamma < \gamma_0$ the origin is asymptotically unstable, and the trajectory $x_\gamma(t)$ is locally pullback attracting.*

Proof By introducing transformation $x = \dfrac{1}{y}$ for Eq. (4.1.19), we see that the solution of the impulsive system (4.1.22) satisfies the following equation, [1, 156, 216],

$$y(t,t_0,y_0) = Y(t,t_0)y_0 + \int_{t_0}^t Y(t,s)q(s)ds + \sum_{t_0 \le \theta_i < t} Y(t,\theta_i+)d_i. \tag{4.1.26}$$

Transforming backward, we have

$$x(t,t_0,x_0) = \frac{1}{Y(t,t_0)x_0^{-1} + \int_{t_0}^t Y(t,s)q(s)ds + \sum_{t_0 \le \theta_i < t} Y(t,\theta_i+)d_i}. \tag{4.1.27}$$

By means of (4.1.20), one can see that asymptotic behavior of (4.1.27) is to depend on sign of γ.

Consider the case $\gamma < 0$. From Eq. (4.1.27) and relation (4.1.20), it follows that $x(t, t_0, x_0) \to 0$ as $t_0 \to -\infty$ for any $x_0 \neq 0$ as long as $x(\xi, t_0, x_0)$ exists for all $\xi \in [t_0, t]$.

For $x_0 > 0$, it is sufficient to show that

$$Y(\xi, t_0)x_0^{-1} + \int_{t_0}^{\xi} Y(\xi, s)q(s)ds + \sum_{t_0 \leq \theta_i < \xi} Y(\xi, \theta_i+)d_i > 0 \qquad (4.1.28)$$

for $\xi \in [t_0, t]$. By means of (4.1.21), inequality (4.1.28) is satisfied provided that

$$Y(\xi, t_0)x_0^{-1} + \int_{T^-}^{\xi} Y(\xi, s)q(s)ds + \sum_{T^- \leq \theta_i < \xi} Y(\xi, \theta_i+)d_i > 0 \qquad (4.1.29)$$

for $\xi \in [T^-, t]$. Because of assumption (4.1.20), for t_0 small enough $Y(\xi, t_0)$ is bounded below on $(-\infty, T^-]$. Thus, (4.1.28) is satisfied if

$$x_0 < \frac{\inf_{t_0 \leq T^-} Y(\xi, t_0)}{\sup_{\xi \in [T^-, t]} \left| \int_{T^-}^{\xi} Y(\xi, s)q(s)ds + \sum_{T^- \leq \theta_i < \xi} Y(\xi, \theta_i+)d_i \right|}. \qquad (4.1.30)$$

For $x_0 < 0$ the argument requires condition (4.1.25), which implies that there exists a μ_t such that

$$\frac{Y(\xi, t_0)}{\int_{t_0}^{\xi} Y(\xi, s)q(s)ds + \sum_{t_0 \leq \theta_i < \xi} Y(\xi, \theta_i+)d_i} \geq \frac{m_\gamma}{2} \qquad (4.1.31)$$

for all $t_0 \leq \mu_t$. Now, it is sufficient to show that

$$Y(\xi, t_0)x_0^{-1} + \int_{t_0}^{\xi} Y(\xi, s)q(s)ds + \sum_{t_0 \leq \theta_i < \xi} Y(\xi, \theta_i+)d_i < 0 \qquad (4.1.32)$$

for $\xi \in [t_0, t]$. Denote $I(t_0, \xi) = \int_{t_0}^{\xi} Y(\xi, s)q(s)ds + \sum_{t_0 \leq \theta_i < \xi} Y(\xi, \theta_i+)d_i$. If $I(t_0, \xi) < 0$, then (4.1.32) is satisfied. If $I(t_0, \xi) > 0$, then we require

$$|x_0| < \frac{Y(\xi, t_0)}{\int_{t_0}^{\xi} Y(\xi, s)q(s)ds + \sum_{t_0 \leq \theta_i < \xi} Y(\xi, \theta_i+)d_i},$$

which has the right-hand side of this expression bounded below by $\frac{m_\gamma}{2}$ using (4.1.31). Therefore, for each t there exists a μ_t such that if $t_0 \leq \mu_t$ and $|x_0|$ is sufficiently small, the solution exists on $[t_0, t]$ and, hence, the origin is locally pullback attracting.

Consider the case when $\gamma > 0$. If $x_0 > 0$, then as $t_0 \to -\infty$ (4.1.27) implies that

$$\lim_{t_0 \to -\infty} x(t, t_0, x_0) = x_\gamma(t) = \frac{1}{\int_{-\infty}^{t} Y(t, s)q(s)ds + \sum_{\theta_i < t} Y(t, \theta_i+)d_i} \qquad (4.1.33)$$

as long as solution exists in the interval $[t_0, t]$. To ensure the existence, it is sufficient to have

$$Y(\xi, t_0)x_0^{-1} + \int_{t_0}^{\xi} Y(\xi, s)q(s)ds + \sum_{t_0 \le \theta_i < \xi} Y(\xi, \theta_i+)d_i > 0 \qquad (4.1.34)$$

for $\xi \in [t_0, t]$. Let us show that (4.1.34) holds if we require $x_0 < (1 + \omega_t)x_\gamma(t_0)$ for some $\omega_t > 0$. Indeed,

$$\begin{aligned}
& Y(\xi, t_0)x_0^{-1} + \int_{t_0}^{\xi} Y(\xi, s)q(s)ds + \sum_{t_0 \le \theta_i < \xi} Y(\xi, \theta_i+)d_i \\
& > \frac{1}{1 + \omega_t} \left\{ \int_{-\infty}^{t_0} Y(\xi, s)q(s)ds + \sum_{\theta_i < t_0} Y(\xi, \theta_i+)d_i \right\} \\
& \quad + \int_{t_0}^{\xi} Y(\xi, s)q(s)ds + \sum_{t_0 \le \theta_i < \xi} Y(\xi, \theta_i+)d_i \qquad (4.1.35) \\
& = \int_{-\infty}^{\xi} Y(\xi, s)q(s)ds + \sum_{\theta_i < \xi} Y(\xi, \theta_i+)d_i \\
& \quad - \frac{\omega_t}{1 + \omega_t} \left\{ \int_{-\infty}^{t_0} Y(\xi, s)q(s)ds + \sum_{\theta_i < t_0} Y(\xi, \theta_i+)d_i \right\} > 0
\end{aligned}$$

for all $t_0 \le \xi \le t$. By (4.1.21), it suffices to show that last expression holds for $\xi \in [T^-, t]$. Choosing $\delta(t) = \omega_t m_\gamma$, it follows that $x_\gamma(t)$ is locally pullback attracting.

Assumption (4.1.24) implies that $0 < \int_{-\infty}^{t} Y(t, s)q(s)ds + \sum_{\theta_i < t} Y(t, \theta_i+)d_i < \infty$. Therefore, from Eq. (4.1.27) and relation (4.1.20), it follows that $x(t, t_0, x_0) \to 0$ as $t \to -\infty$, which implies that the origin is asymptotically unstable.

If $x_0 < 0$, then in order to solution $x(\xi, t_0, x_0)$ not to blow up in a finite time we need

$$Y(\xi, t_0)x_0^{-1} + \int_{t_0}^{\xi} Y(\xi, s)q(s)ds + \sum_{t_0 \le \theta_i < \xi} Y(\xi, \theta_i+)d_i < 0,$$

for all $\xi \in [t_0, t]$. The last relation is satisfied if $I(\xi, t_0) = \int_{t_0}^{\xi} Y(\xi, s)q(s)ds + \sum_{t_0 \le \theta_i < \xi} Y(\xi, \theta_i+)d_i < 0$. If $I(\xi, t_0) > 0$, we choose

$$|x_0| < \frac{Y(\xi, t_0)}{\int_{t_0}^{\xi} Y(\xi, s)q(s)ds + \sum_{t_0 \le \theta_i < \xi} Y(\xi, \theta_i+)d_i},$$

which is bounded from below as it was proven in the case $\gamma < 0$.
This finalizes the proof of the theorem. $\quad\square$

Next, we want to formulate an impulsive extension of the system (4.1.19), which is related to the forward attraction. We assume that the function $q(t)$ and the numbers d_i are asymptotically positive as $t \to \infty$ and $\theta_i \to \infty$, respectively, and the balance condition (4.1.24) is valid. That is,

$$q(t) \geq \overline{b} > 0 \text{ for all } t \geq T^+, \text{ and } d_i \geq \overline{d} > 0 \text{ for all } \theta_i \geq T^+. \qquad (4.1.36)$$

$$0 < m_\gamma \leq x_\gamma(t) = \frac{1}{\int_{-\infty}^t Y(t,s)q(s)ds + \sum_{\theta_i < t} Y(t,\theta_i+)d_i} \leq M_\gamma \qquad (4.1.37)$$

for all $t \in \mathbb{R}, \quad 0 < \gamma < \gamma_0$.

Theorem 4.1.3 *Assume above conditions hold for Eq.(4.1.19). Then, for $-\gamma_0 < \gamma < 0$ the origin is locally forward attracting and for $0 < \gamma < \gamma_0$ the trajectory $x_\gamma(t)$ is locally forward attracting. In addition, if*

$$0 < m_\gamma \leq x_\gamma(t) = \frac{1}{\int_t^\infty Y(t,s)q(s)ds + \sum_{\theta_i < t} Y(t,\theta_i+)d_i} \leq M_\gamma \qquad (4.1.38)$$

for all $t \in \mathbb{R}, \gamma < 0$, then for $-\gamma_0 < \gamma < 0$ the trajectory $x_\gamma(t)$ is both asymptotically unstable and locally pullback repelling.

Proof If $\gamma < 0$, the origin is locally forward attracting when x_0 is sufficiently small, since condition (4.1.36) implies that

$$\inf_{t \geq t_0} \left\{ \int_{t_0}^t Y(t,s)q(s)ds + \sum_{t_0 \leq \theta_i < t} Y(t,\theta_i+)d_i \right\} > -\infty. \qquad (4.1.39)$$

If $\gamma > 0$, the trajectory $x_\gamma(t)$ is locally forward attracting, which easily follows from the following relation.

$$\left(\frac{1}{x(t)} - \frac{1}{x_\gamma(t)} \right) = Y(t,t_0) \left(\frac{1}{x_0} - \frac{1}{x_\gamma(t_0)} \right). \qquad (4.1.40)$$

Therefore,

$$|x(t) - x_\gamma(t)| = \frac{x_\gamma(t)x(t)}{x_\gamma(t_0)x_0} e^{\left(\frac{-\int_{t_0}^t p(u)du + \sum_{t_0 \leq \theta_i < t} \ln c_i}{t - t_0} \right)(t - t_0)} |x_0 - x_\gamma(t_0)|. \qquad (4.1.41)$$

Using the balance condition (4.1.37) with $x_0 > 0$ implies that

$$x(t) = \frac{1}{Y(t, t_0)x_0^{-1} + \int_{t_0}^t Y(t, s)q(s)ds + \sum_{t_0 \le \theta_i < t} Y(t, \theta_i+)d_i}$$

$$\le M_\gamma \frac{\int_{-\infty}^t Y(t, s)q(s)ds + \sum_{\theta_i < t} Y(t, \theta_i+)d_i}{Y(t, t_0)x_0^{-1} + \int_{t_0}^t Y(t, s)q(s)ds + \sum_{t_0 \le \theta_i < t} Y(t, \theta_i+)d_i} \tag{4.1.42}$$

$$= M_\gamma \frac{Y(t, t_0)x_\gamma^{-1}(t_0) + \int_{t_0}^t Y(t, s)q(s)ds + \sum_{t_0 \le \theta_i < t} Y(t, \theta_i+)d_i}{Y(t, t_0)x_0^{-1} + \int_{t_0}^t Y(t, s)q(s)ds + \sum_{t_0 \le \theta_i < t} Y(t, \theta_i+)d_i}.$$

Condition (4.1.36) guarantees that for t sufficiently large the integral and the sum in the numerator and denominator are positive. So, from the last expression it follows that

$$\limsup_{t \to \infty} x(t) \le M_\gamma max \left\{ 1, \frac{x_0}{x_\gamma(t_0)} \right\}.$$

Therefore, any solution with $x_0 > 0$ is bounded as $t \to \infty$. Hence, from (4.1.41) it follows that $x_\gamma(t)$ is forward attracting as long as solutions do not blow up in a finite time. Next, we show that solution exists for $x_0 < (1 + \omega_{t_0})x_\gamma(t_0)$.

$$Y(t, t_0)x_0^{-1} + \int_{t_0}^t Y(t, s)q(s)ds + \sum_{t_0 \le \theta_i < t} Y(t, \theta_i+)d_i$$

$$> \frac{1}{1 + \omega_{t_0}} \left\{ \int_{-\infty}^{t_0} Y(t, s)q(s)ds + \sum_{\theta_i < t_0} Y(t, \theta_i+)d_i \right\}$$

$$+ \int_{t_0}^t Y(t, s)q(s)ds + \sum_{t_0 \le \theta_i < t} Y(t, \theta_i+)d_i \tag{4.1.43}$$

$$= \int_{-\infty}^t Y(t, s)q(s)ds + \sum_{\theta_i < t} Y(t, \theta_i)d_i$$

$$- \frac{\omega_{t_0}}{1 + \omega_{t_0}} \left\{ \int_{-\infty}^{t_0} Y(t, s)q(s)ds + \sum_{\theta_i < t_0} Y(t, \theta_i+)d_i \right\}.$$

The last expression is positive for sufficiently small ω_{t_0} because of the assumption (4.1.36). Therefore, $x_\gamma(t)$ is locally forward attracting.

Under the final assumption (4.1.38), the results follow by making the transformations

$$\gamma \mapsto -\gamma, \quad x \mapsto -x, \quad \theta \mapsto -\theta \text{ and } t \mapsto -t.$$

This finalizes the proof. $\square$

Example 4.1.2 Let $p(t) \equiv a$, $c_i \equiv c$, and $\theta_i = ih$ for the system (4.1.19) with $h > 0$. That is,

$$x' = ax - q(t)x^2,$$
$$\Delta x|_{t=ih} = -x + \frac{x}{c+d_i x}. \tag{4.1.44}$$

Then, $\gamma = a - \frac{1}{h}\ln c$. By means of $y = \frac{1}{x}$, the system (4.1.17) is converted to the linear impulsive system

$$
\begin{aligned}
\dot{y} &= -ay + q(t), \\
\Delta y|_{t=ih} &= (c-1)y + d_i.
\end{aligned}
\tag{4.1.45}
$$

Asymptotic behavior of (4.1.45) is to depend on the sign of γ, and results of Theorems 4.1.2 and 4.1.3 are true for the system (4.1.44). If $c = 1$ and $d_i = 0$, then $\gamma = a$ and there is no equation of jumps in the system (4.1.44).

4.2 Impulsive Bernoulli Equations: The Transcritical and the Pitchfork Bifurcations

In this section, we can study the existence of the bounded solutions and asymptotic behavior of impulsive Bernoulli equations. Moreover, we generalize that nonautonomous pitchfork and transcritical bifurcation scenarios are investigated in the previous section. Illustrative examples with numerical simulations are given to support our theoretical results.

4.2.1 Introduction and Preliminaries

The Bernoulli equations constitute an important class of nonlinear differential equations. In this section, we shall introduce a new type of impulsive equations. We say that an impulsive equation is of the Bernoulli type if it is reducible to an equation, which is linear and nonhomogeneous in both its components, differential and impulsive. Thus, it is essentially nonlinear not only in its differential equation, but also in its impulsive part. It is important to note that the equation, which is under discussion in this section, is obtained not by a simple adding of an impulsive expression to the differential Bernoulli equation. Moreover, to the best of our knowledge, there is no study which deals with discontinuous Bernoulli equations at all.

It is only in the recent decades, there have been intensive developments on time-dependent differential equations. Local theory of dynamical systems is concerned with asymptotic behavior of a fixed point or a periodic solution. However, in nonautonomous dynamical systems it is usually hard to find a fixed point or a periodic solution. Indeed, in many cases they fail even to exist. Therefore, the notion of fixed points is generically endure as bounded solutions in the theory of time-varying dynamical systems. There are abstract formulation of a nonautonomous dynamical systems as a new concept of nonautonomous attractors which are called pullback attractors [81, 83, 139, 155, 204, 227]. We investigate appearances and disappearances of bounded solutions that are stable and unstable in the pullback and forward sense. In particular, it was possible to study bifurcation analysis in nonautonomous systems with pullback attractors [75, 138, 152, 154]. In previous section, we have studied nonautonomous transcritical and pitchfork bifurcations in impulsive systems. In the present section, we introduce a new and the most general impulsive Bernoulli equation and discuss bifurcation analysis of these equations. The main equation under investigation is the following impulsive system,

$$
\begin{aligned}
x' &= p(t)x - q(t)x^n, \\
\Delta x|_{t=\theta_i} &= -x + \frac{x}{(c_i + d_i x^{n-1})^{\frac{1}{n-1}}},
\end{aligned}
\tag{4.2.46}
$$

where the functions $p, q : \mathbb{R} \to \mathbb{R}$ are continuous, the sequence of real numbers $\{\theta_i\}$, $i \in \mathbb{N}$, is such that there exist two real numbers $\underline{\theta}$ and $\overline{\theta}$ satisfying $\underline{\theta} \le \theta_{i+1} - \theta_i \le \overline{\theta}$, $\Delta x|_{t=\theta_i} := x(\theta_i+) - x(\theta_i)$ and $x(\theta_i+) = \lim_{t \to \theta_i^+} x(t)$. The system (4.2.46) is nonlinear not only in its differential equation part but in its impulsive part also. It consists of the Bernoulli equation and nonlinear impulsive one such that under the Bernoulli transformation, $y = x^{1-n}$, it is reduced to the following linear impulsive nonhomogeneous equation,

$$
\begin{aligned}
\dot{y} &= (1 - n)p(t)y + (n - 1)q(t), \\
\Delta y|_{t=\theta_i} &= (c_i - 1)y + d_i.
\end{aligned}
$$

This is the reason why we call (4.2.46) to be the impulsive Bernoulli equation. Moreover, the results obtained for the system (4.2.46) are also interpreted for the following continuous Bernoulli equation,

$$
x' = p(t)x - q(t)x^n,
\tag{4.2.47}
$$

where the functions $p, q : \mathbb{R} \to \mathbb{R}$ are continuous. Thus, new results are accomplished for (4.2.47). Let $x(t, t_0, x_0)$ be solution of (4.2.46) or (4.2.47). In this section, we deal with scalar differential equations such that $x(t, t_0, x_0)$ is continuable on $\mathbb{R}$. Solutions are unique both forward and backward in time. In the previous section, we have considered the system (4.2.46) for $n = 2$ and $n = 3$. However, we did not state forward asymptotic analysis for the case $n = 3$. In this section, we state results for forward and pullback asymptotic analyses and n is allowed to be an arbitrary natural number. Moreover, we obtain conditions for (4.2.46) and (4.2.47) to have nontrivial bounded solutions on $\mathbb{R}$. Pullback asymptotic analysis of Eq. (4.2.47) with $p(t) = const.$ and $n = 3$ has been carried out by Caraballo and Langa in [75] and Langa et al., in [152]. In [154], the authors considered (4.2.47) for $n = 2$ and $p(t) = \lambda a(t)$, where different bifurcation analyses are studied depending on the sign of the λ. In this section, we want to emphasize that we obtain different bifurcation scenarios for (4.2.46) which depend on $\gamma = \limsup\limits_{t-s \to \infty} \frac{\int_s^t (1-n)p(u)du + \sum_{s \le \theta_i \le t} \ln c_i}{t-s}$ and for (4.2.47) depend on $\gamma = \limsup\limits_{t-s \to \infty} \frac{\int_s^t (1-n)p(u)du}{t-s}$. Thus, the discontinuous system (4.2.46) satisfies the bifurcation conditions for the wider class of functions $p(t)$ than for continuous system (4.2.47). In other words, the bifurcation is cost by change of the exponents of a solution. This approach is premised for the first time in the literature in our papers [30, 32]. We continue with this idea in the present section and significantly improve the results obtained in the previous section. A theory of nonautonomous bifurcations in a Banach space is treated in terms of exponential dichotomy in a series of remarkable papers [198–200].

4.2.2 Bounded Solutions

In this section, we study the existence of a bounded solution of (4.2.46). It is easy to see that $x = 0$ is the trivial bounded solution of (4.2.46). In what follows, we are interested in the solutions which are bounded and different from zero. For this purpose, we shall need the following conditions.

(C1)   There exist positive real numbers m and M such that $0 < m \le q(t) \le M$ for all $t \in \mathbb{R}$;

(C2)   There exists positive real number and L such that $0 \le d_i \le L$ for all $i \in \mathbb{Z}$.

By means of the transformation $y = x^{1-n}$, (4.2.46) is reduced to the following linear impulsive system,

$$\begin{aligned} \dot{y} &= (1-n)p(t)y + (n-1)q(t), \\ \Delta y|_{t=\theta_i} &= (c_i - 1)y + d_i. \end{aligned} \tag{4.2.48}$$

Let $\Psi(t, s)$ be the fundamental matrix of (4.2.48). One can find that

$$\Psi(t, s) = e^{\int_s^t (1-n)p(u)du} \prod_{s \le \theta_i \le t} c_i = e^{\int_s^t (1-n)p(u)du + \sum_{s \le \theta_i \le t} \ln c_i}.$$

Denote $\gamma = \limsup\limits_{t-s \to \infty} \dfrac{\int_s^t (1-n)p(u)du + \sum_{s \le \theta_i \le t} \ln c_i}{t-s}$. One can guarantee that there exist two positive numbers k and K such that

$$ke^{\gamma(t-s)} \le \|\Psi(t, s)\| \le Ke^{\gamma(t-s)}, \quad s \le t. \tag{4.2.49}$$

Lemma 4.2.1 *If (C1)–(C2) are satisfied, then (4.2.46) admits a nontrivial bounded on $\mathbb{R}$ solutions $\tilde{x}(t)$ which satisfy the following equations*

$$\tilde{x}^{n-1}(t) = \frac{1}{\int_{-\infty}^t \Psi(t, s)(n-1)q(s)ds + \sum_{\theta_i < t} \Psi(t, \theta_i+)d_i}, \quad \text{if } \gamma < 0,$$

$$\tilde{x}^{n-1}(t) = -\frac{1}{\int_t^\infty \Psi(t, s)(n-1)q(s)ds - \sum_{t \le \theta_i < \infty} \Psi(t, \theta_i+)d_i}, \quad \text{if } \gamma > 0.$$

Proof Consider $\gamma < 0$. It suffices to show that $\widetilde{y}(t) = \int_{-\infty}^{t} \Psi(t, s)(n - 1)q(s)ds + \sum_{\theta_i < t} \Psi(t, \theta_i+)d_i$ is a bounded solution of (4.2.48). Let us verify that $\widetilde{y}(t)$ satisfies (4.2.48).

$$
\begin{aligned}
\dot{\widetilde{y}}(t) &= (1 - n)p(t) \int_{-\infty}^{t} \Psi(t, s)(n - 1)q(s)ds \\
&\quad + (n - 1)\Psi(t, t)q(t) + (1 - n)p(t) \sum_{\theta_i < t} \Psi(t, \theta_i+)d_i \\
&= (1 - n)p(t) \left\{ \int_{-\infty}^{t} \Psi(t, s)(n - 1)q(s)ds + \sum_{\theta_i < t} \Psi(t, \theta_i+)d_i \right\} \\
&\quad + (n - 1)q(t) \\
&= (1 - n)p(t)\widetilde{y}(t) + (n - 1)q(t).
\end{aligned}
$$

To show that $\widetilde{y}(t)$ satisfies the equation jumps, we note for fixed j it is true that $\Psi(\theta_j+, s) - \Psi(\theta_j, s) = (c_j - 1)\Psi(\theta_j, s)$. Thus, $\Psi(\theta_j+, s) = c_j\Psi(\theta_j, s)$.

$$
\begin{aligned}
\Delta\widetilde{y}(t)|_{t=\theta_j} &= \widetilde{y}(\theta_j+) - \widetilde{y}(\theta_j) \\
&= \int_{-\infty}^{\theta_j+} \Psi(\theta_j+, s)(n - 1)q(s)ds + \sum_{\theta_i < \theta_j+} \Psi(\theta_j+, \theta_j+)d_j \\
&\quad - \int_{-\infty}^{\theta_j} \Psi(\theta_j, s)(n - 1)q(s)ds - \sum_{\theta_i < \theta_j} \Psi(\theta_j, \theta_j+)d_j \\
&= c_j \int_{-\infty}^{\theta_j} \Psi(\theta_j, s)(n - 1)q(s)ds - \int_{-\infty}^{\theta_j} \Psi(\theta_j, s)(n - 1)q(s)ds + d_j \\
&\quad + \sum_{\theta_i < \theta_j} c_j\Psi(\theta_j, \theta_j+)d_j - \sum_{\theta_i < \theta_j} \Psi(\theta_j, \theta_j+)d_j \\
&= (c_j - 1) \left\{ \int_{-\infty}^{\theta_j} \Psi(\theta_j, s)(n - 1)q(s)ds + \sum_{\theta_i < \theta_j} \Psi(\theta_j, \theta_j+)d_j \right\} + d_j \\
&= (c_j - 1)\widetilde{y}(\theta_j) + d_j.
\end{aligned}
$$

Next, we show that $\widetilde{y}(t)$ is bounded and separated from zero.

$$
\begin{aligned}
0 < \frac{mk(n - 1)}{-\gamma} &\leq \|\widetilde{y}(t)\| \leq \frac{MK(n - 1)}{-\gamma} + LK \sum_{\theta_i < t} e^{\gamma(t-\theta_i)} \\
&\leq \frac{MK(n - 1)}{-\gamma} + LK \sum_{i=0}^{\infty} e^{i\gamma\underline{\theta}} = K\left(\frac{M(n - 1)}{-\gamma} + \frac{L}{1 - e^{\gamma\underline{\theta}}}\right) < \infty.
\end{aligned}
$$

Now consider $\gamma > 0$. Similarly, it suffices to show that $\widetilde{y}(t) = -\int_{t}^{\infty} \Psi(t, s)(n - 1)q(s)ds - \sum_{t \leq \theta_i < \infty} \Psi(t, \theta_i+)d_i$ is a bounded solution of (4.2.48). Let us verify that $\widetilde{y}(t)$ satisfies (4.2.48).

$$\dot{\tilde{y}}(t) = -(1-n)p(t)\int_t^\infty \Psi(t,s)(n-1)q(s)ds + (n-1)\Psi(t,t)q(t)$$

$$- (1-n)p(t)\sum_{t\le\theta_i<\infty}\Psi(t,\theta_i+)d_i$$

$$= (1-n)p(t)\left\{-\int_t^\infty \Psi(t,s)(n-1)q(s)ds - \sum_{t\le\theta_i<\infty}\Psi(t,\theta_i+)d_i\right\}$$

$$+ (n-1)q(t)$$

$$= (1-n)p(t)\tilde{y}(t) + (n-1)q(t).$$

To show that $\tilde{y}(t)$ satisfies the equation of jumps, we note for fixed j, it is true that $\Psi(\theta_j+,s) - \Psi(\theta_j,s) = (c_j-1)\Psi(\theta_j,s)$. Thus, $\Psi(\theta_j+,s) = c_j\Psi(\theta_j,s)$.

$$\Delta\tilde{y}(t)|_{t=\theta_j} = \tilde{y}(\theta_j+) - \tilde{y}(\theta_j)$$

$$= -\int_{\theta_j+}^\infty \Psi(\theta_j+,s)(n-1)q(s)ds - \sum_{\theta_j+\le\theta_i}\Psi(\theta_j+,\theta_j+)d_j$$

$$+ \int_{\theta_j}^\infty \Psi(\theta_j,s)(n-1)q(s)ds + \sum_{\theta_j\le\theta_i}\Psi(\theta_j,\theta_j+)d_j$$

$$= -c_j\int_{\theta_j}^\infty \Psi(\theta_j,s)(n-1)q(s)ds + \int_{\theta_j}^\infty \Psi(\theta_j,s)(n-1)q(s)ds + d_j$$

$$- \sum_{\theta_j\le\theta_i}c_j\Psi(\theta_j,\theta_j+)d_j + \sum_{\theta_j\le\theta_i}\Psi(\theta_j,\theta_j+)d_j$$

$$= (c_j-1)\left\{-\int_{\theta_j}^\infty \Psi(\theta_j,s)(n-1)q(s)ds - \sum_{\theta_j\le\theta_i}\Psi(\theta_j,\theta_j+)d_j\right\} + d_j$$

$$= (c_j-1)\tilde{y}(\theta_j) + d_j.$$

Next, we show that $\tilde{y}(t)$ is bounded and separated from zero.

$$0 < \frac{mk(n-1)}{\gamma} \le ||\tilde{y}(t)|| \le \frac{MK(n-1)}{\gamma} + LK\sum_{t\le\theta_i}e^{\gamma(t-\theta_i)}$$

$$\le \frac{MK(n-1)}{\gamma} + LK\sum_{i=0}^\infty e^{-i\gamma\bar{\theta}} = K\left(\frac{M(n-1)}{\gamma} + \frac{L}{1-e^{\gamma\bar{\theta}}}\right) < \infty.$$

Therefore, $\tilde{y}(t)$ is bounded and by (C1) it is separated from zero. The lemma is proved. □

Finally, let us show that $\tilde{y}(t)$ is a unique solution of (4.2.48). Assume on the contrary that there exists bounded solution $y_1(t)$ different from $\tilde{y}(t)$. Then, $w_0(t) =: \tilde{y}(t) - y_1(t)$ is a bounded solution of the following linear impulsive system

$$\dot{w} = (1-n)p(t)y,$$

$$\Delta w|_{t=\theta_i} = (c_i-1)w. \tag{4.2.50}$$

But (4.2.50) admits only the trivial solution bounded on $\mathbb{R}$. Therefore, $\widetilde{y}(t) = y_1(t)$ and it implies that all bounded solutions which is different from zero have to satisfy the following equations,

$$\widetilde{x}^{1-n}(t) = \int_{-\infty}^{t} \Psi(t,s)(n-1)q(s)ds + \sum_{\theta_i < t} \Psi(t,\theta_i+)d_i, \quad \text{if } \gamma < 0,$$

$$\widetilde{x}^{1-n}(t) = -\int_{t}^{\infty} \Psi(t,s)(n-1)q(s)ds - \sum_{t \leq \theta_i < \infty} \Psi(t,\theta_i+)d_i, \quad \text{if } \gamma > 0.$$

Thus, one can see that if n is even, then there is unique nontrivial bounded solution. If n is odd, then there are two nontrivial bounded solutions. In what follows, we have different bifurcation scenarios depending on the parity of n. In the next sections, we deal with pitchfork and transcritical bifurcations, respectively.

4.2.3 The Pitchfork Bifurcation

Consider (4.2.46) for $n = 2m + 1$. That is,

$$\begin{aligned} x' &= p(t)x - q(t)x^{2m+1}, \\ \Delta x|_{t=\theta_i} &= -x + \frac{x}{\left(c_i + d_i x^{2m}\right)^{\frac{1}{2m}}}, \end{aligned} \tag{4.2.51}$$

where $p, q \in PC(\mathbb{R}, \theta)$, $m \in \mathbb{N}$ and $c_i, d_i \in \mathbb{R}^+$ for all $i \in \mathbb{Z}$.

Theorem 4.2.1 *Suppose that (C1)–(C2) are fulfilled for (4.2.51). Then, for $\gamma > 0$ the trivial solution is asymptotically pullback and forward stable whereas the nontrivial bounded solutions $\widetilde{x}(t)$ are asymptotically unstable, and for $\gamma < 0$ the trivial solution is asymptotically unstable and the nontrivial bounded solutions are asymptotically pullback and forward stable.*

Proof One can find that the solution of (4.2.51) satisfies the following equation, [1, 156, 216],

$$x^{2m}(t, t_0, x_0) = \frac{1}{\Psi(t,t_0)x_0^{-2m} + 2\int_{t_0}^{t} \Psi(t,s)mq(s)ds + \sum_{t_0 \leq \theta_i < t} \Psi(t,\theta_i+)d_i}. \tag{4.2.52}$$

In the previous section, we have shown that (4.2.51) admits the trivial solution and two bounded solutions which satisfy the following equations

$$
\widetilde{x}(t) = \begin{cases} \pm \left(\dfrac{1}{2 \int_{-\infty}^{t} \Psi(t,s) mq(s) ds + \sum_{\theta_i < t} \Psi(t, \theta_i +) d_i} \right)^{\frac{1}{2m}}, & \text{if } \gamma < 0 \\[4mm] \pm \left(-\dfrac{1}{2 \int_{t}^{\infty} \Psi(t,s) mq(s) ds + \sum_{t \le \theta_i < \infty} \Psi(t, \theta_i +) d_i} \right)^{\frac{1}{2m}}, & \text{if } \gamma > 0 \end{cases}.
$$

One can see that asymptotic behavior of (4.2.51) depends on γ. We start with the case $\gamma > 0$. From (4.2.52), it follows that $x^{2m}(t, t_0, x_0) \to 0$ as $t_0 \to -\infty$ as well as $t \to \infty$. So, $x(t, t_0, x_0) \to 0$ as $t_0 \to -\infty$ and as $t \to \infty$, replying that all solutions are attracted both forward and pullback to the point $\{0\}$.

To show that the nontrivial bounded solutions $\widetilde{x}(t)$ are asymptotically unstable notice that

$$
x^{-2m}(t) - \widetilde{x}^{-2m}(t) = \Psi(t, t_0) \left(x_0^{-2m} - \widetilde{x}^{-2m}(t_0) \right). \tag{4.2.53}
$$

From the last expression, it follows that $x(t)$ converges to $\widetilde{x}(t)$ as $t \to -\infty$ whenever $\|x_0\| < \|\widetilde{x}(t_0)\|$.

If $\gamma < 0$, we notice that the expression (4.2.53) holds. Thus, one can see that $x(t)$ converges to $\widetilde{x}(t)$ both forward and pullback whenever $\|x_0\| < \|\widetilde{x}(t_0)\|$. To show that the origin is asymptotically unstable we rewrite the expression (4.2.53) as follows:

$$
x^{2m}(t) = \frac{1}{\Psi(t, t_0) \left(x_0^{-2m} - \widetilde{x}^{-2m}(t_0) \right) + \widetilde{x}^{-2m}(t)},
$$

which implies that $x(t)$ converges to 0 as $t \to -\infty$ whenever $\|x_0\| < \|\widetilde{x}(t_0)\|$. The theorem is proved. $\square$

Remark 4.2.1 In the similar manner, it can be easily shown that the results of Theorem 4.2.1 hold for the following system:

$$
x' = p(t)x - q(t)x^{2m+1},
$$
$$
\Delta x|_{t=\theta_i} = -x - \frac{x}{\left(c_i + d_i x^{2m} \right)^{\frac{1}{2m}}}.
$$

We obtain the similar results for the following equation:

$$
x' = p(t)x - q(t)x^{2m+1} \tag{4.2.54}
$$

where p *and* q are continuous functions. For this particular case, we have that $\gamma = \limsup_{t-s \to \infty} \frac{\int_s^t (1-n)p(u) du}{t-s}$. If (C1) satisfied, one can show that the nontrivial bounded solutions satisfy the following equations,

$$\tilde{x}(t) = \begin{cases} \pm\left(\dfrac{1}{2\int_{-\infty}^{t}\Psi(t,s)mq(s)ds}\right)^{\frac{1}{2m}}, & \text{if } \gamma < 0 \\[4mm] \pm\left(-\dfrac{1}{2\int_{t}^{\infty}\Psi(t,s)mq(s)ds}\right)^{\frac{1}{2m}}, & \text{if } \gamma > 0 \end{cases}.$$

Theorem 4.2.2 *Suppose that (C1) is fulfilled for (4.2.54). Then, for $\gamma > 0$ the trivial solution is asymptotically pullback and forward stable whereas the nontrivial bounded solutions $\tilde{x}(t)$ are asymptotically unstable and for $\gamma < 0$ the trivial solution is asymptotically unstable and the nontrivial bounded solutions are asymptotically pullback and forward stable.*

We omit the proof since it is similar to that of Theorem 4.2.1.

4.2.4 The Transcritical Bifurcation

In this section, we consider (4.2.46) for $n = 2m$. That is,

$$\begin{aligned} x' &= p(t)x - q(t)x^{2m}, \\ \Delta x|_{t=\theta_i} &= -x + \frac{x}{(c_i+d_ix^{2m-1})^{\frac{1}{2m-1}}}, \end{aligned} \qquad (4.2.55)$$

where $c_i, d_i \in \mathbb{R}^+, i \in \mathbb{Z}, p, q \in PC(\mathbb{R}, \theta)$.

Theorem 4.2.3 *Suppose that (C1)–(C2) are fulfilled for (4.2.55). Then, for $\gamma > 0$ the trivial solution is asymptotically forward and pullback stable and for $\gamma < 0$ the trivial solution is asymptotically unstable and the nontrivial bounded solution is forward and pullback stable.*

Proof One can find that the solution of (4.2.55) satisfies the following equation, [1, 156, 216],

$$x^{2m-1}(t, t_0, x_0) = \frac{1}{\Psi(t,t_0)x_0^{-2m+1} + \int_{t_0}^{t}\Psi(t,s)(2m-1)q(s)ds + \sum_{t_0\leq\theta_i<t}\Psi(t,\theta_i+)d_i} \cdot \qquad (4.2.56)$$

In previous section, we have shown that (4.2.55) admits the trivial solution and the nontrivial bounded solution which satisfy the following equations

$$\widetilde{x}(t)= \begin{cases} \left(\dfrac{1}{\int_{-\infty}^{t} \Psi(t,s)(2m-1)q(s)ds + \sum_{\theta_i < t} \Psi(t,\theta_i+)d_i} \right)^{\frac{1}{2m-1}}, & \text{if } \gamma < 0 \\[4mm] \left(-\dfrac{1}{\int_{t}^{\infty} \Psi(t,s)(2m-1)q(s)ds + \sum_{t \le \theta_i < \infty} \Psi(t,\theta_i+)d_i} \right)^{\frac{1}{2m-1}}, & \text{if } \gamma > 0 \end{cases}.$$

As in the previous section, it is clear that asymptotic behavior of (4.2.55) depends on the sign of γ. Consider the case $\gamma > 0$. From the Eq. (4.2.56), it follows that $x(t, t_0, x_0) \to 0$ as $t_0 \to -\infty$ and as $t \to \infty$ as long as $x(\xi, t_0, x_0)$ exists for all $\xi \in [t_0, t]$. If $x_0 > 0$, observe that $\Psi(\xi, t_0)x_0^{-2m+1} + \int_{t_0}^{\xi} \Psi(\xi, s)(2m-1)q(s)ds + \sum_{t_0 \le \theta_i < \xi} \Psi(\xi, \theta_i+)d_i > 0$,
for $\xi \in [t_0, t]$. Thus, $x(\xi, t_0, x_0)$ exists for all $\xi \in [t_0, t]$ and does not blow up as $t_0 \to -\infty$ and as $t \to \infty$.

If $x_0 < 0$, to ensure the existence of the solution $x(\xi, t_0, x_0)$ it is sufficient to show that $\Psi(\xi, t_0)x_0^{-2m+1} + \int_{t_0}^{\xi} \Psi(\xi, s)(2m-1)q(s)ds + \sum_{t_0 \le \theta_i < \xi} \Psi(\xi, \theta_i+)d_i < 0$, for $\xi \in [t_0, t]$. Since $\int_{t_0}^{\xi} \Psi(\xi, s)(2m-1)q(s)ds + \sum_{t_0 \le \theta_i < \xi} \Psi(\xi, \theta_i+)d_i > 0$, we require

$$|x_0| < \left(\frac{\Psi(\xi, t_0)}{\int_{t_0}^{\xi} \Psi(\xi, s)(2m-1)q(s)ds + \sum_{t_0 \le \theta_i < \xi} \Psi(\xi, \theta_i+)d_i} \right)^{\frac{1}{2m-1}}.$$

However, we need to show that right-hand side of the last inequality is bounded from below. One can find that

$$\frac{\Psi(\xi, t_0)}{\int_{t_0}^{\xi} \Psi(\xi, s)(2m-1)q(s)ds + \sum_{t_0 \le \theta_i < \xi} \Psi(\xi, \theta_i+)d_i}$$
$$= \frac{1}{-\widetilde{x}^{-2m+1}(t_0) + \Psi^{-1}(\xi, t_0)\widetilde{x}^{-2m+1}(t)}.$$

It is easy to see that the last expression is bounded from below since $\widetilde{x}(t)$ is bounded and $\Psi^{-1}(\xi, t_0)$ is bounded for small enough t_0 or for large enough ξ.

Finally, we consider the case $\gamma < 0$. To show that the trivial solution is asymptotically unstable notice that

$$x^{2m-1}(t) = \frac{1}{\Psi(t, t_0)\left(x_0^{-2m+1} - \widetilde{x}^{-2m+1}(t_0)\right) + \widetilde{x}^{-2m+1}(t)}. \qquad (4.2.57)$$

From the last expression, it follows that $x(t)$ converges to 0 as $t \to -\infty$ for all $0 < x_0 < \widetilde{x}(t_0)$.

It remains to show that $\widetilde{x}(t)$ is forward and pullback stable. If $x_0 > 0$, then it is clear that $\Psi(\xi, t_0)x_0^{-2m+1} + \int_{t_0}^{\xi} \Psi(\xi, s)(2m-1)q(s)ds + \sum_{t_0 \le \theta_i < \xi} \Psi(\xi, \theta_i+)d_i > 0$, for $\xi \in [t_0, t]$. Thus, the solution $x(\xi, t_0, x_0)$ exists for all $\xi \in [t_0, t]$ and (4.2.57) implies that $\widetilde{x}(t)$ is forward and pullback stable for all $0 < x_0 < \widetilde{x}(t_0)$.

If $x_0 < 0$, then to ensure the existence of the solution $x(\xi, t_0, x_0)$ it is sufficient to show that $\Psi(\xi, t_0)x_0^{-2m+1} + \int_{t_0}^{\xi} \Psi(\xi, s)(2m-1)q(s)ds + \sum_{t_0 \leq \theta_i < \xi} \Psi(\xi, \theta_i+)d_i < 0$, for $\xi \in [t_0, t]$. Since $\int_{t_0}^{\xi} \Psi(\xi, s)(2m-1)q(s)ds + \sum_{t_0 \leq \theta_i < \xi} \Psi(\xi, \theta_i+)d_i > 0$, we require

$$|x_0| < \left(\frac{\Psi(\xi, t_0)}{\int_{t_0}^{\xi} \Psi(\xi, s)(2m-1)q(s)ds + \sum_{t_0 \leq \theta_i < \xi} \Psi(\xi, \theta_i+)d_i} \right)^{\frac{1}{2m-1}}.$$

The right-hand side of the last inequality is bounded from below because the following relation holds.

$$\frac{\Psi(\xi, t_0)}{\int_{t_0}^{\xi} \Psi(\xi, s)(2m-1)q(s)ds + \sum_{t_0 \leq \theta_i < \xi} \Psi(\xi, \theta_i+)d_i}$$
$$= \frac{1}{-\tilde{x}^{-2m+1}(t_0) + \Psi^{-1}(\xi, t_0)\tilde{x}^{-2m+1}(t)}.$$

Proof of the theorem is finalized. $\square$

Remark 4.2.2 In the previous section, we have considered (4.2.55) for $m = 1$ and required asymptotic positivity for $q(t)$ and d_i instead of the conditions (C1)–(C2). Namely, we assumed that there exist positive constants $\bar{q}$ and $\bar{d}$ such that $q(t) \geq \bar{q}$ for all $t \leq T$, and $d_i \geq \bar{d}$ for all $\theta_i \leq T$. Moreover, to ensure the existence of solution we required the balance conditions. That is to say we assumed that there exists $\gamma_0 > 0$ such that $0 < m < \tilde{x}(t) < M$, for all $0 < \gamma < \gamma_0$, and $\liminf_{t_0 \to -\infty} \frac{\Psi(t,t_0)}{\int_{t_0}^{t} \Psi(t,s)(2m-1)q(s)ds + \sum_{t_0 \leq \theta_i < t} \Psi(t,\theta_i+)d_i} \geq m > 0$ for all $-\gamma_0 < \gamma < 0$ hold. However, in the present section we show that the conditions (C1)–(C2) are enough to ensure the existence of the solution.

Finally, we state the similar results for the following equation.

$$x' = p(t)x - q(t)x^{2m}, \qquad (4.2.58)$$

where p and q are continuous functions. For this particular case, we have that $\gamma = \limsup_{t-s \to \infty} \frac{\int_s^t (1-n)p(u)du}{t-s}$. If (C1) satisfied, one can show that the nontrivial bounded solutions satisfy the following equations,

$$\tilde{x}(t) = \begin{cases} \left(\dfrac{1}{\int_{-\infty}^{t} \Psi(t, s)(2m-1)q(s)ds} \right)^{\frac{1}{2m-1}}, & \text{if } \gamma < 0 \\[4ex] \left(-\dfrac{1}{\int_{t}^{\infty} \Psi(t, s)(2m-1)q(s)ds} \right)^{\frac{1}{2m-1}}, & \text{if } \gamma > 0 \end{cases}.$$

Theorem 4.2.4 *Suppose that (C1) is fulfilled for (4.2.58). Then, for $\gamma > 0$ the trivial solution is asymptotically forward and pullback stable and for $\gamma < 0$ the trivial solution is asymptotically unstable and the nontrivial bounded solution is forward and pullback stable.*

We omit the proof since it is similar to that of Theorem 4.2.3.

4.2.5   Illustrative Examples

In this section, to illustrate theoretical results of Theorem 4.2.1 we consider two examples.

Example 4.2.1 Let us consider the following system,

$$
\begin{aligned}
x' &= (6 + 2.5 \sin(t^2))x - (18 + 3.5 \cos(1 + \tfrac{t^2}{5}))x^7, \\
\Delta x|_{t=i} &= -x + \frac{x}{\left(\frac{i^2}{i^2+3} + \frac{10x^6}{1+i^2}\right)^{\frac{1}{6}}},
\end{aligned}
\tag{4.2.59}
$$

where $p(t) = 6 + 2.5\sin(t^2)$, $q(t) = 18 + 3.5\cos(1 + \tfrac{t^2}{5})$, $\theta_i = i, i \in \mathbb{N}, c_i = \frac{i^2}{i^2+3}$, $d_i = \frac{10}{1+i^2}$ and $n = 7$. We check that all conditions of Theorem 4.2.1 are satisfied with $m = 14.5$, $M = 21.5$, and $L = 5$. However, we do not verify all consequences of Theorem 4.2.1. One can see that $\gamma = \limsup\limits_{t-s \to \infty} \frac{\int_s^t (-36 - 15\sin(u^2))du + \sum_{s \leq i \leq t} \ln \frac{i^2}{i^2+3}}{t-s} <$ 0. Thus, Theorem 4.2.1 guarantees that (4.2.59) has nontrivial bounded solutions which satisfy equations $\tilde{x}(t) = \pm \left(\frac{1}{6\int_\infty^t \Psi(t,s)(18+3.5\cos(1+\frac{s^2}{5}))ds + \sum_{s \leq i < \xi} \Psi(\xi,i) + \frac{10}{1+i^2}} \right)^{\frac{1}{6}}$,

where $\Psi(t,s) = e^{-6\int_s^t (6+2.5\sin(s^2))ds} \prod\limits_{s \leq i \leq t} \frac{i^2}{i^2+3}$. Figure 4.1 reveals that all solutions starting near the origin diverge from the origin and converge to the nontrivial bounded solutions $\pm\tilde{x}(t)$. Therefore, the origin is asymptotically unstable and the bounded solutions are forward and pullback stable as expressed in the numerical simulations. Moreover, from the simulations it is seen that the nontrivial bounded solution satisfies the inequality $0.6 \leq ||\tilde{x}(t)|| \leq 1$.

Example 4.2.2 We consider the following system,

$$
\begin{aligned}
x' &= -\left(1.01 + \sin\left(5 + \tfrac{t^3}{5}\right)\right)x - \left(0.21 + 0.2\cos\left(1 + \tfrac{t^2}{5}\right)\right)x^7, \\
\Delta x|_{t=i} &= -x + \frac{x}{\left(\frac{i^2+3}{i^2} + \frac{10x^6}{1+i^2}\right)^{\frac{1}{6}}},
\end{aligned}
\tag{4.2.60}
$$

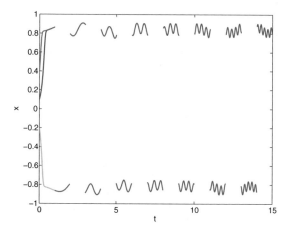

Fig. 4.1 Asymptotic behavior of (4.2.59) for $t \in [0, 15]$. In the figure, the *green color* corresponds to the solution with initial value $x_0 = -0.2$, the *blue color* corresponds to the solution with initial value $x_0 = 0.1$, and the *red color* corresponds to the solution with initial value $x_0 = 0.4$. One can see that all solutions which start in the neighborhood of the origin diverge from the origin and converge to the nontrivial bounded solutions $\pm \widetilde{x}(t)$, which cannot see through the simulations

where $p(t) = -1.01 - \sin(5 + \frac{t^3}{5})$, $q(t) = 0.21 + 0.2\cos(1 + \frac{t^2}{5})$, $\theta_i = i, i \in \mathbb{N}$, $c_i = \frac{i^2+3}{i^2}$, $d_i = \frac{10}{1+i^2}$ and $n = 7$. We check that all conditions of Theorem 4.2.1 are satisfied with $m = 0.01$, $M = 0.41$ and $L = 5$. However, we do not verify all consequences of Theorem 4.2.1. Note that for this example one can confirm that

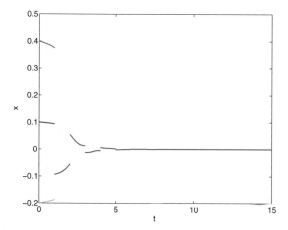

Fig. 4.2 Asymptotic behavior of (4.2.60) for $t \in [0, 15]$. In the figure, the *green color* corresponds to the solution with initial value $x_0 = -0.2$, the *blue color* corresponds to the solution with initial value $x_0 = 0.1$, and the *red color* corresponds to the solution with initial value $x_0 = 0.4$. One can see that all solutions which start in the neighborhood of the origin eventually converge to the origin

$\gamma = \limsup\limits_{t-s \to \infty} \dfrac{\int_s^t (6.06+6\sin(5+\frac{u^3}{5}))du+\sum_{s\le i\le t} \ln \frac{i^2+3}{i^2}}{t-s} > 0$. Figure 4.2 reveals that all solutions starting near the origin eventually converge to the origin. Thus, the origin is forward and pullback stable as expressed in the numerical simulation.

4.3 Notes

There are qualitative studies on asymptotic behavior of impulsive systems [1, 4, 37, 156, 216]. There are also many studied which deal with bifurcation theory either in autonomous differential equations [1, 6, 41] or periodic equations with fixed moments of impulses [66, 109, 111]. However, differential equations with fixed moments of impulses are naturally nonautonomous differential equations. Consequently, one cannot construct the theory similar to autonomous systems of ordinary differential equations. Thus, in order to achieve results on fixed moments, it is crucial to extend idea of pullback attraction to impulsive systems for nonautonomous differential equations. Although the theory of impulsive differential equations is very developed nowadays, there are no results concerning analogues in [61, 75, 81, 97, 98, 139, 142, 151, 155, 227]. This appears to be due to the absence of papers concerning concrete systems analyzing the existence of nonautonomous bifurcations. It is hoped that present chapter fills this gap. The main novelty of the current study is to construct nonautonomous bifurcation theory for impulsive systems with appropriate definitions of pullback attracting sets. This is the very first step toward the bifurcation of nonautonomous differential equations with impulses.

The pitchfork and the transcritical bifurcations are considered for nonautonomous impulsive differential equations. Explicitly solvable models with the specific equations of jumps have been considered. This allowed us to categorize one-dimensional bifurcations in impulsive systems.

This theory could be developed in many ways. One can consider impulsive analogues for the pitchfork bifurcation and corresponding impulsive analogue for the transcritical bifurcation without finding explicit solution similarly to that done in [204]. Nonautonomous saddle-node bifurcation remains unconsidered even for one-dimensional impulsive systems. Finally, general theory of bifurcation in higher-dimensional systems with impulses has to be developed.

In the last section, it is the first time the impulsive Bernoulli equation has been studied. It is clearly seen that in the second part of the chapter we have obtained results that are more general than those in the first section. This chapter provides new sufficient conditions guaranteeing the existence of the nontrivial bounded solutions. Moreover, both forward and pullback asymptotic behavior of the trivial and the nontrivial bounded solutions are studied. Different nonautonomous bifurcation scenarios depending on the asymptotic behavior of these solutions are obtained. The results obtained in [30, 32] constitute the main part of this chapter.

Chapter 5
Nonautonomous Bifurcations in Nonsolvable Impulsive Systems

5.1 The Transcritical and the Pitchfork Bifurcations

In this section, we consider impulsive analogues of nonautonomous transcritical and pitchfork bifurcations in the systems which cannot be solved explicitly. We extend the theorem on asymptotic properties of the quasilinear impulsive systems which was considered by Samoilenko and Perestyuk [216] to the entire time.

5.1.1 Introduction

In the previous chapter, we studied nonautonomous bifurcations which are solvable by means of Bernoulli transformation. In particular, the concept of the pullback and forward attracting sets was used to analyze nonautonomous bifurcations which depend on the properties of the system in the past and future, respectively. However, most of nonlinear discontinuous models are not of a Bernoulli type. In this chapter, we are concerned with the most general type of equations, which are naturally cannot be solved. Thus, we cannot expect to find a bounded or a periodic solution as in the previous chapter. Instead, we deal with asymptotic behavior of an equilibrium point. Thus, we need different approach than one in the previous chapter in order to describe bifurcation analysis.

A system undergoes bifurcation if a particular solution of a differential equation gain or loss attractivity when parameter varies. Therefore, there is a strong relation between notions of attractiveness/repulsiveness of a solution and bifurcation theory. In this chapter, we implement this idea to study various bifurcation scenarios and focus on systems with discontinuity. There are qualitative studies on asymptotic properties of the quasilinear impulsive systems of differential equations [1, 4, 37, 156, 216]. In Chap. 4, we studied impulsive extensions of the nonautonomous pitchfork and transcritical bifurcation in the systems which are explicitly solvable models. The

© Springer Nature Singapore Pte Ltd. and Higher Education Press 2017
M. Akhmet and A. Kashkynbayev, *Bifurcation in Autonomous
and Nonautonomous Differential Equations with Discontinuities*,
Nonlinear Physical Science, DOI 10.1007/978-981-10-3180-9_5

main novelty of this chapter is to study analogues of nonautonomous pitchfork and transcritical bifurcations in scalar impulsive systems which depend on properties of the system on entire time, i.e., past and future time. Moreover, we extend the results on stability of quasilinear systems based on first-order approximation obtained by Samoilenko and Perestyuk [216] to entire time.

5.1.2 Preliminaries

We denote by $\mathbb{R}$ the set of all real numbers, $\mathbb{Z}$ the set of integers and write $\mathbb{R}_k^- := [-\infty, k)$ and $\mathbb{R}_k^+ := [k, \infty)$ for a given $k \in \mathbb{R}$. In this section, we introduce concepts of attractive and repulsive solutions, which are used to analyze asymptotic behavior of impulsive systems. This section is concerned with systems of the type

$$
\begin{aligned}
x' &= f(t, x), \\
\Delta x|_{t=\theta_i} &= J_i(x),
\end{aligned}
\tag{5.1.1}
$$

where $\Delta x|_{t=\theta_i} := x(\theta_i+) - x(\theta_i)$, $x(\theta_i+) = \lim_{t \to \theta_i^+} x(t)$. The system (5.1.1) is defined on the set $\Omega = I \times \mathbb{A} \times G$ where $G \subseteq \mathbb{R}^n$, I is the interval of the form $I = \mathbb{R}$, $I = \mathbb{R}_k^-$ or $I = \mathbb{R}_k^+$, respectively. θ is a nonempty sequence with the set of indexes $\mathbb{A}$ such that $|\theta_i| \to \infty$ as $|i| \to \infty$. Let $\phi(t, t_0, x_0)$ be solution of (5.1.1) which is continuable and unique on I.

Denote $PC(\mathbb{R}, \theta)$ space of piecewise left continuous functions with discontinuity of the first kind at points in θ. In this chapter, the Euclidean norm $|| \cdot ||$ and Hausdorff semi-distance between nonempty set X and Y as $d(X, Y) = \sup_{x \in X} \inf_{y \in Y} d(x, y)$ are used. For arbitrary ε-neighborhood of some point $x_0 \in \mathbb{R}^n$ we write $B_\varepsilon(x_0) = \{x \in \mathbb{R}^n : ||x - x_0|| < \varepsilon\}$ and for arbitrary nonempty set $X \subset \mathbb{R}^n$ we define $\phi(t, t_0, X) := \bigcup_{x_0 \in X} \phi(t, t_0, x_0)$. A graph of function $g : A \to B$ is defined as $graph g = \{(a, b) \in A \times B : g(a) = b\}$.

A set $\mathcal{N} \subset I \times \mathbb{R}^n$ is called nonautonomous set if the set $\mathcal{N}(t) := \{x \in \mathbb{R}^n : (t, x) \in \mathcal{N}\}$, called as t−fibers, is not empty for all $t \in I$. $\mathcal{N}$ is said to be compact if all t−fibers are compact and $\mathcal{N}$ is said to be invariant if it satisfies $\phi(t, t_0, \mathcal{N}(t_0)) = \mathcal{N}(t)$ for all $t, t_0 \in I$.

Asymptotic properties of discontinuous dynamics and continuous one are the same. In what follows, we use definitions of attractivity and repulsivity without any changes form [204].

Definition 5.1.1 Let $\psi : \mathbb{R} \to \mathbb{R}^n$ be a solution of the system (5.1.1).

- A compact and invariant nonautonomous set $\mathcal{G}$ is all-time attractor if there exists an $\varepsilon > 0$ such that

$$
\lim_{\substack{t \to \infty \\ t_0 \in \mathbb{R}}} \sup d(\phi(t + t_0, t_0, B_\varepsilon(\mathcal{G}(t_0))), \mathcal{G}(t + t_0)) = 0.
$$

All-time attraction radius of $\mathscr{G}$, denoted by $\mathscr{A}_{\mathscr{G}}^{\pm}$, is the supremum of all positive ε which satisfy the above relation;
- If *graphψ* is an all-time attractor, then $\psi(t)$ is called all-time attractive;
- A compact and invariant nonautonomous set $\mathscr{H}$ is all-time repeller if there exists an $\varepsilon > 0$ such that

$$\lim_{\substack{t \to \infty \\ t_0 \in \mathbb{R}}} \sup d(\phi(t_0 - t, t_0, B_\varepsilon(\mathscr{H}(t_0))), \mathscr{H}(t_0 - t)) = 0.$$

All-time repulsion radius of $\mathscr{H}$, denoted by $\mathscr{R}_{\mathscr{H}}^{\pm}$, is the supremum of all positive ε which satisfy the above relation;
- If *graphψ* is an all-time repeller, then ψ is called all-time repulsive.

In Chap. 4, we studied nonautonomous bifurcation patterns in the pullback and forward sense. In what follows to examine asymptotic analysis of the systems that depend in the past we define past attractivity and repulsivity. One can confirm that a past attractor is a local pullback attractor [82].

Definition 5.1.2 Let $\psi : \mathbb{R}_k^- \to \mathbb{R}^n$ be a solution of the system (5.1.1).

- A compact and invariant nonautonomous set $\mathscr{G}$ is past attractor if there exists an $\varepsilon > 0$ such that

$$\lim_{t \to \infty} d(\phi(t_0, t_0 - t, B_\varepsilon(\mathscr{G}(t_0 - t))), \mathscr{G}(t_0)) = 0 \ \text{ for all } \ t_0 \in \mathbb{R}_k^-.$$

The past attraction radius of $\mathscr{G}$, denoted by $\mathscr{A}_{\mathscr{G}}^-$, is the supremum of all positive ε which satisfy the above relation;
- If *graphψ* a past attractor, then $\psi(t)$ is called past attractive;
- A compact and invariant nonautonomous set $\mathscr{H}$ is past repeller if there exists an $\varepsilon > 0$ such that

$$\lim_{t \to \infty} d(\phi(t_0 - t, t_0, B_\varepsilon(\mathscr{H}(t_0))), \mathscr{H}(t_0 - t)) = 0 \ \text{ for all } \ t_0 \in \mathbb{R}_k^-.$$

Past repulsion radius of $\mathscr{H}$, denoted by $\mathscr{R}_{\mathscr{H}}^-$, is the supremum of all positive ε such that there exists a $\widehat{k} \in \mathbb{R}_k^-$ with

$$\lim_{t \to \infty} d(\phi(t_0 - t, t_0, B_\varepsilon(\mathscr{H}(t_0))), \mathscr{H}(t_0 - t)) = 0 \ \text{ for all } \ t_0 \leq \widehat{k};$$

- If *graphψ* is a past repeller, then $\psi(t)$ is called past repulsive.

Definition 5.1.3 Let $\psi : \mathbb{R}_k^+ \to \mathbb{R}^n$ be a solution of the system (5.1.1).

- A compact and invariant nonautonomous set $\mathscr{G}$ is future attractor if there exists an $\varepsilon > 0$ such that

$$\lim_{t \to \infty} d(\phi(t + t_0, t_0, B_\varepsilon(\mathscr{G}(t_0))), \mathscr{G}(t + t_0)) = 0 \ \text{ for all } \ t_0 \in \mathbb{R}_k^+.$$

Future attraction radius of $\mathscr{G}$, denoted by $\mathscr{A}_{\mathscr{G}}^{+}$, is the supremum of all positive ε such that there exists a $\widehat{k} \in \mathbb{R}_{k}^{+}$ with

$$\lim_{t \to \infty} d(\phi(t + t_0, t_0, B_\varepsilon(\mathscr{G}(t_0))), \mathscr{G}(t + t_0)) = 0 \text{ for all } t_0 \geq \widehat{k};$$

- If *graph* ψ is a future attractor, then $\psi(t)$ is called future attractive;
- A compact and invariant nonautonomous set $\mathscr{H}$ is called future repeller if there exists an $\varepsilon > 0$ with

$$\lim_{t \to \infty} d(\phi(t_0, t_0 + t, B_\varepsilon(\mathscr{H}(t + t_0))), \mathscr{H}(t_0)) = 0 \text{ for all } t_0 \in \mathbb{R}_{k}^{+}.$$

Future repulsion radius of $\mathscr{H}$, denoted by $\mathscr{R}_{\mathscr{H}}^{+}$, is the supremum of all positive ε with the above relation;
- If *graph* ψ is a future repeller, then $\psi(t)$ is called future repulsive.

From definitions given above, it follows that if a solution is future attractive, then it is uniformly asymptotically stable. Moreover, every all-time attractor/repeller is both a past attractor/repeller and a future attractor/repeller.

5.1.3   Attractivity and Repulsivity in a Linear Impulsive Nonhomogeneous Systems

By means of definitions given above, we analyze attractivity and repulsivity in linear nonhomogeneous systems which are important in the stability analysis of solutions of nonlinear impulsive systems with fixed moments of impulses. We consider the system with interval I of the form $\mathbb{R}$, $\mathbb{R}_{k}^{-}$, or $\mathbb{R}_{k}^{+}$, respectively, and let

$$\begin{aligned} x' &= A(t)x + F(t, x), \\ \Delta x|_{t=\theta_i} &= B_i x + I_i(x), \end{aligned} \tag{5.1.2}$$

where $A \in PC(I, \theta)$, matrices B_i satisfy $det(B_i + I) \neq 0$, $F : I \times G \to \mathbb{R}^n$ and $I : \mathbb{A} \times G \to \mathbb{R}^n$. An infinite sequence θ_i satisfies $|\theta_i| \to \infty$ as $|i| \to \infty$. It is assumed that there exist positive constants $\underline{\theta}$ and $\overline{\theta}$ such that $\underline{\theta} \leq \theta_{i+1} - \theta_i \leq \overline{\theta}$. Denote $\phi(t, t_0, x_0)$ as the solution of (5.1.2) and $\Psi(t, s)$ as the fundamental matrix of the following system

$$\begin{aligned} x' &= A(t)x, \\ \Delta x|_{t=\theta_i} &= B_i x. \end{aligned} \tag{5.1.3}$$

Theorem 5.1.1 *If there exist $\alpha < 0$, $K \geq 1$, and $\delta > 0$ such that*

$$||\Psi(t, s)|| \leq Ke^{\alpha(t-s)} \text{ for all } t \geq s,$$

and the functions $F(t, x)$ and $I_i(x)$ are Lipschitzian, i.e., there exists a positive number l such that

$$||F(t, x)|| \leq l||x||, \quad ||I_i(x)|| \leq l||x|| \tag{5.1.4}$$

for all $t \in I$, $i \in \mathbb{A}$ and $||x|| < h$, $h > 0$. Then,

$$||\phi(t, t_0, x_0)|| \leq \delta e^{(\alpha + Kl + \frac{1}{\underline{\theta}} \ln(1 + Kl))(t - t_0)} \quad \text{for all} \ \ t, t_0 \in I \ \ \text{with} \ \ t \geq t_0,$$

i.e., for sufficiently small values of l, the origin is all-time attractive.

Proof An equivalent integral equation of the system (5.1.2) can be written as [1, 216]:

$$\phi(t, t_0, x_0) = \Psi(t, t_0)x_0 + \int_{t_0}^{t} \Psi(t, s)F(s, \phi(s, t_0, x_0))ds$$

$$+ \sum_{t_0 \leq \theta_i < t} \Psi(t, \theta_i)I_i(\phi(\theta_i, t_0, x_0))$$

for all $t \geq t_0$. By using inequalities in (5.1.4), we get

$$||\phi(t, t_0, x_0)|| \leq Ke^{\alpha(t - t_0)}||x_0|| + \int_{t_0}^{t} Ke^{\alpha(t - s)}l||\phi(s, t_0, x_0)||ds$$

$$+ \sum_{t_0 \leq \theta_i < t} Ke^{\alpha(t - \theta_i)}l||\phi(\theta_i, t_0, x_0)||$$

for all $t \geq t_0$ is fulfilled. The last expression can be rewritten as

$$e^{-\alpha t}||\phi(t, t_0, x_0)|| \leq Ke^{-\alpha t_0}||x_0|| + \int_{t_0}^{t} Kle^{-\alpha s}||\phi(s, t_0, x_0)||ds$$

$$+ \sum_{t_0 \leq \theta_i < t} Kle^{-\alpha \theta_i}||\phi(\theta_i, t_0, x_0)||$$

for all $t \geq t_0$. Hence, by Gronwall–Bellman lemma for piecewise continuous functions ([1, 216]) it follows that

$$||\phi(t, t_0, x_0)|| \leq Ke^{(\alpha + Kl)(t - t_0)}(1 + Kl)^{i[t_0, t)}||x_0|| \quad \text{for all} \ \ t \geq t_0. \tag{5.1.5}$$

By means of the inequality $\theta_{i+1} - \theta_i \geq \underline{\theta}$, one can see that

$$||\phi(t, t_0, x_0)|| \leq Ke^{(\alpha + Kl + \frac{1}{\underline{\theta}} \ln(1 + Kl))(t - t_0)}||x_0|| \quad \text{for all} \ \ t \geq t_0.$$

If l is small enough that

$$\alpha + Kl + \frac{1}{\underline{\theta}} \ln(1 + Kl) < 0 \ \text{for} \ \alpha < 0,$$

then the required result follows by choosing $\delta = Kh$. □

Theorem 5.1.2 *If there exist $\alpha > 0$, $K \geq 1$, and $\delta > 0$ such that*

$$||\Psi(t,s)|| \leq Ke^{\alpha(t-s)} \ \text{for all} \ t \leq s,$$

and the functions $F(t,x)$ and $I_i(x)$ are Lipschitzian, i.e., there exists a positive number l such that

$$||F(t,x)|| \leq l||x||, \quad ||I_i(x)|| \leq l||x|| \tag{5.1.6}$$

for all $t \in I$, $i \in \mathbb{A}$ and $||x|| < h$, $h > 0$. Then,

$$||\phi(t,t_0,x_0)|| \leq \delta e^{(\alpha - Kl + \frac{1}{\underline{\theta}} \ln(1-Kl))(t-t_0)} \ \text{for all} \ t, t_0 \in I \ \text{with} \ t \leq t_0,$$

i.e., for sufficiently small values of l, the origin is all-time repulsive.

Proof An equivalent integral equation of the system (5.1.2) can be written as [1, 216]:

$$\phi(t,t_0,x_0) = \Psi(t,t_0)x_0 + \int_{t_0}^{t} \Psi(t,s)F(s,\phi(s,t_0,x_0))ds$$
$$- \sum_{t \leq \theta_i < t_0} \Psi(t,\theta_i)I_i(\phi(\theta_i,t_0,x_0))$$

for all $t \leq t_0$. By using inequalities in (5.1.4) we get

$$||\phi(t,t_0,x_0)|| \leq Ke^{\alpha(t-t_0)}||x_0|| + \int_{t_0}^{t} Ke^{\alpha(t-s)}l||\phi(s,t_0,x_0)||ds$$
$$+ \sum_{t \leq \theta_i < t_0} Ke^{\alpha(t-\theta_i)}l||\phi(\theta_i,t_0,x_0)||$$

for all $t \leq t_0$ is fulfilled. The last expression can be rewritten as

$$e^{-\alpha t}||\phi(t,t_0,x_0)|| \leq Ke^{-\alpha t_0}||x_0|| + \int_{t_0}^{t} Kle^{-\alpha s}||\phi(s,t_0,x_0)||ds$$
$$+ \sum_{t \leq \theta_i < t_0} Kle^{-\alpha \theta_i}||\phi(\theta_i,t_0,x_0)||$$

for all $t \leq t_0$. Gronwall–Bellman lemma for piecewise continuous functions ([1, 216]) can be applied since l can be chosen such that $Kl < 1$. Thus,

$$||\phi(t, t_0, x_0)|| \leq K e^{(\alpha - Kl)(t - t_0)} (1 - Kl)^{-i[t, t_0]} ||x_0|| \text{ for all } t \leq t_0. \tag{5.1.7}$$

By means of the inequality $\theta_{i+1} - \theta_i \geq \underline{\theta}$, one can see that

$$||\phi(t, t_0, x_0)|| \leq K e^{(\alpha - Kl + \frac{1}{\theta} \ln(1 - Kl))(t - t_0)} ||x_0|| \text{ for all } t \leq t_0.$$

If l is small enough that

$$\alpha - Kl + \frac{1}{\underline{\theta}} \ln(1 - Kl) > 0 \text{ for } \alpha > 0,$$

then the required results follow if we choose $\delta = Kh$ since $||x_0|| < h.$ $\square$

5.1.4   The Transcritical Bifurcation

In this section, we study impulsive analogue of the nonautonomous transcritical bifurcation. Let $x_- < 0 < x_+$ and $\mu_- < \mu_+$ be real numbers and I be interval of the form $\mathbb{R}$, $\mathbb{R}_k^-$, or $\mathbb{R}_k^+$, respectively. Consider the system

$$\begin{aligned} x' &= p(t, \mu)x + q(t, \mu)x^2 + r(t, x, \mu), \\ \Delta x|_{t = \theta_i} &= c_i(\mu)x + d_i(\mu)x^2 + e_i(x, \mu), \end{aligned} \tag{5.1.8}$$

with piecewise continuous functions $p : I \times (\mu_-, \mu_+) \to \mathbb{R}$, $q : I \times (\mu_-, \mu_+) \to \mathbb{R}$ and $r : I \times (x_-, x_+) \times (\mu_-, \mu_+) \to \mathbb{R}$ satisfying $r(t, 0, \mu) = 0$. $c : \mathbb{A} \times (\mu_-, \mu_+) \to \mathbb{R}$, $d : \mathbb{A} \times (\mu_-, \mu_+) \to \mathbb{R}$ and $e : \mathbb{A} \times (x_-, x_+) \times (\mu_-, \mu_+) \to \mathbb{R}$ with $c_i(\mu) \neq -1$ and $e_i(0, \mu) = 0$. An infinite sequence θ_i satisfies $|\theta_i| \to \infty$ as $|i| \to \infty$. It is assumed that there exist positive constants $\underline{\theta}$ and $\overline{\theta}$ such that $\underline{\theta} \leq \theta_{i+1} - \theta_i \leq \overline{\theta}$. Let $\Psi_\mu(t, s)$ be the fundamental matrix of the following linear system.

$$\begin{aligned} x' &= p(t, \mu)x, \\ \Delta x|_{t = \theta_i} &= c_i(\mu)x. \end{aligned} \tag{5.1.9}$$

Assume that there exists $\mu_0 \in (\mu_-, \mu_+)$ such that there are two functions $\alpha_1, \alpha_2 : (\mu_-, \mu_+) \to \mathbb{R}$ which are either both monotone increasing or both monotone decreasing satisfying $\lim_{\mu \to \mu_0} \alpha_1(\mu) = \lim_{\mu \to \mu_0} \alpha_2(\mu) = 0$ and $K \geq 1$ such that

$$||\Psi_\mu(t, s)|| \leq K e^{\alpha_1(\mu)(t - s)} \text{ for all } \mu \in (\mu_-, \mu_+) \text{ and } t, s \in I \text{ with } t \geq s,$$

$$||\Psi_\mu(t, s)|| \leq K e^{\alpha_2(\mu)(t - s)} \text{ for all } \mu \in (\mu_-, \mu_+) \text{ and } t, s \in I \text{ with } t \leq s.$$

The functions q and d_i satisfy one of the following conditions.

$$0 < \liminf_{\mu \to \mu_0} \inf_{t \in I} q(t, \mu) \le \limsup_{\mu \to \mu_0} \sup_{t \in I} q(t, \mu) < \infty,$$

$$0 < \liminf_{\mu \to \mu_0} \inf_{i \in \mathbb{A}} d_i(\mu) \le \limsup_{\mu \to \mu_0} \sup_{i \in \mathbb{A}} d_i(\mu) < \infty \tag{5.1.10}$$

or

$$-\infty < \liminf_{\mu \to \mu_0} \inf_{t \in I} q(t, \mu) \le \limsup_{\mu \to \mu_0} \sup_{t \in I} q(t, \mu) < 0,$$

$$-\infty < \liminf_{\mu \to \mu_0} \inf_{i \in \mathbb{A}} d_i(\mu) \le \limsup_{\mu \to \mu_0} \sup_{i \in \mathbb{A}} d_i(\mu) < 0. \tag{5.1.11}$$

The functions r and e_i satisfy the following conditions.

$$\lim_{x \to 0} \sup_{\mu \in (\mu_0 - |x|, \mu_0 + |x|)} \sup_{t \in I} \frac{|r(t, x, \mu)|}{|x|^2} = 0,$$

$$\lim_{x \to 0} \sup_{\mu \in (\mu_0 - |x|, \mu_0 + |x|)} \sup_{i \in \mathbb{A}} \frac{|e_i(x, \mu)|}{|x|^2} = 0 \tag{5.1.12}$$

and there exists sufficiently small $l > 0$ such that

$$|r(t, x, \mu)| < l|x|, \quad |e_i(x, \mu)| < l|x|, \tag{5.1.13}$$

for all $\mu \in (\mu_-, \mu_+), t \in I, i \in \mathbb{A}$ and $x \in (x_-, x_+)$.

Theorem 5.1.3 *Assume that above conditions hold for the system (5.1.8). Then there exists $\widehat{\mu}_- < 0 < \widehat{\mu}_+$ such that*

- *If the functions α_1 and α_2 are monotone increasing, the origin is all-time attractive for $\mu \in (\widehat{\mu}_-, \mu_0)$ and all-time repulsive for $\mu \in (\mu_0, \widehat{\mu}_+)$. The following relations hold true.*

$$\lim_{\mu \to \mu_0^-} \mathscr{A}_0^\mu = 0 \quad and \quad \lim_{\mu \to \mu_0^+} \mathscr{R}_0^\mu = 0; \tag{5.1.14}$$

- *If the functions α_1 and α_2 are monotone decreasing, the origin is all-time repulsive for $\mu \in (\widehat{\mu}_-, \mu_0)$ and all-time attractive for $\mu \in (\mu_0, \widehat{\mu}_+)$. The following relations hold true.*

$$\lim_{\mu \to \mu_0^+} \mathscr{A}_0^\mu = 0 \quad and \quad \lim_{\mu \to \mu_0^-} \mathscr{R}_0^\mu = 0; \tag{5.1.15}$$

Hence, in both of the above cases the system (5.1.8) possesses an all-time bifurcation.

Proof Let ϕ_μ be the general solution of the system (5.1.8). We may consider the case (5.1.10) since the functions α_1 and α_2 are monotone increasing. Choose $\widehat{\mu}_- < \mu_0 < \widehat{\mu}_+$ such that

$$0 < \inf_{\mu \in (\widehat{\mu}_-, \widehat{\mu}_+), t \in I} q(t, \mu) \le \sup_{\mu \in (\widehat{\mu}_-, \widehat{\mu}_+), t \in I} q(t, \mu) < \infty,$$

$$0 < \inf_{\mu \in (\widehat{\mu}_-, \widehat{\mu}_+), i \in \mathbb{A}} d_i(\mu) \le \sup_{\mu \in (\widehat{\mu}_-, \widehat{\mu}_+), i \in \mathbb{A}} d_i(\mu) < \infty. \tag{5.1.16}$$

By means of (5.1.13) and (5.1.16), one can see that Theorems 5.1.1 and 5.1.2 can be applied. Thus, we get attractivity and repulsivity of the origin as it was required to show in the theorem. It remains to show that relations (5.1.14) and (5.1.15) hold. Let us assume on the contrary that

$$\gamma = \limsup_{\mu \to \mu_0^-} \mathscr{A}_0^\mu > 0.$$

By means of (5.1.12) and (5.1.16) one can show that there exists $\widetilde{\mu} \in (\widehat{\mu}_-, \mu_0)$, $x_0 \in (0, \gamma)$ and $J \in (0, \frac{x_0}{4K})$ such that

$$q(t, \mu)x^2 + r(t, x, \mu) > J \quad \text{and} \quad d_i(\mu)x^2 + e_i(x, \mu) > J \tag{5.1.17}$$

for all $t \in I$, $i \in \mathbb{A}$, $\mu \in (\widetilde{\mu}_-, \mu_0)$ and $x_0 \in \left[\frac{x_0}{2K^2}, x_0\right]$. Next, fix $\widehat{\mu} \in (\widetilde{\mu}_-, \mu_0)$ such that $\mathscr{A}_0^{\widehat{\mu}} > x_0$ and $\alpha_2(\widehat{\mu}) \geq \alpha := -\frac{2KJ}{x_0} > -\frac{1}{2}$ so that $\phi_{\widehat{\mu}}(t, t_0, x_0)$ is attracted to the origin. Set $\psi(t) = \phi_{\widehat{\mu}}(t, t_0, x_0)$. Then, there exists $\tau \in I$, $\tau > t_0$, such that $\psi(\tau) \leq \frac{x_0}{2K^2}$. We choose minimal τ which satisfy this property. In other words, $\psi(\tau) > \frac{x_0}{2K^2}$ for all $t \in [t_0, \tau)$. Moreover, choose $t_1 \in [t_0, \tau)$ such that

$$\psi(t_1) = \frac{x_0}{2K} \quad \text{and} \quad \psi(t) \in \left(\frac{x_0}{2K^2}, x_0\right] \quad \text{for all } t \in [t_1, \tau).$$

We write integral equation of the system (5.1.8) at $t = \tau$ for fixed $\widehat{\mu}$ which start at point t_1.

$$\begin{aligned}
\psi(\tau) =& \Psi_{\widehat{\mu}}(\tau, t_1)\psi(t_1) + \int_{t_1}^\tau \Psi_{\widehat{\mu}}(\tau, s)\left(b(s, \widehat{\mu})(\psi(s))^2 + r(s, \psi(s), \widehat{\mu})\right) ds \\
&+ \sum_{t_1 \leq \theta_i < \tau} \Psi_{\widehat{\mu}}(\tau, \theta_i)\left(d_i(\widehat{\mu})(\psi(\theta_i))^2 + e_i(\psi(\theta_i), \widehat{\mu})\right) \\
&> \frac{x_0}{2K^2}e^{\alpha(\tau - t_1)} + \frac{J}{K}\int_{t_1}^\tau e^{\alpha(\tau - s)} ds \\
&= e^{\alpha(\tau - t_1)}\left(\frac{x_0}{2K^2} + \frac{J}{K\alpha}\right) - \frac{J}{K\alpha} = \frac{x_0}{2K^2}
\end{aligned}$$

which is a contradiction and we arrive at $\lim_{\mu \to \mu_0^-} \mathscr{A}_0^\mu = 0$. In the similar fashion, one can show that $\lim_{\mu \to \mu_0^+} \mathscr{R}_0^\mu = 0$. We omit the proof of the second part since it be verified in the similar manner. This finalizes the proof of theorem. $\square$

5.1.5 The Pitchfork Bifurcation

In this section, we study impulsive analogue of the nonautonomous pitchfork bifurcation. Let $x_- < 0 < x_+$ and $\mu_- < \mu_+$ be real numbers and I be interval of the form

$\mathbb{R}, \mathbb{R}_k^-$ or $\mathbb{R}_k^+$, respectively. Consider the system

$$
\begin{aligned}
x' &= p(t, \mu)x + q(t, \mu)x^3 + r(t, x, \mu), \\
\Delta x|_{t=\theta_i} &= c_i(\mu)x + d_i(\mu)x^3 + e_i(x, \mu),
\end{aligned}
\tag{5.1.18}
$$

with piecewise continuous functions $p : I \times (\mu_-, \mu_+) \to \mathbb{R}, q : I \times (\mu_-, \mu_+) \to \mathbb{R}$ and $r : I \times (x_-, x_+) \times (\mu_-, \mu_+) \to \mathbb{R}$ satisfying $r(t, 0, \mu) = 0$. $c : \mathbb{A} \times (\mu_-, \mu_+) \to \mathbb{R}, d : \mathbb{A} \times (\mu_-, \mu_+) \to \mathbb{R}$ and $e : \mathbb{A} \times (x_-, x_+) \times (\mu_-, \mu_+) \to \mathbb{R}$ with $c_i(\mu) \neq -1$ and $e_i(0, \mu) = 0$. An infinite sequence θ_i satisfies $|\theta_i| \to \infty$ as $|i| \to \infty$. It is assumed that there exist positive constants $\underline{\theta}$ and $\overline{\theta}$ such that $\underline{\theta} \leq \theta_{i+1} - \theta_i \leq \overline{\theta}$. Let $\Psi_\mu(t, s)$ be the fundamental matrix of the following linear system.

$$
\begin{aligned}
x' &= p(t, \mu)x, \\
\Delta x|_{t=\theta_i} &= c_i(\mu)x.
\end{aligned}
\tag{5.1.19}
$$

Assume that there exists $\mu_0 \in (\mu_-, \mu_+)$ such that there are two functions $\alpha_1, \alpha_2 : (\mu_-, \mu_+) \to \mathbb{R}$ which are either both monotone increasing or both monotone decreasing satisfying $\lim_{\mu \to \mu_0} \alpha_1(\mu) = \lim_{\mu \to \mu_0} \alpha_2(\mu) = 0$ and $K \geq 1$ such that

$$
\|\Psi_\mu(t, s)\| \leq K e^{\alpha_1(\mu)(t-s)} \text{ for all } \mu \in (\mu_-, \mu_+) \text{ and } t, s \in I \text{ with } t \geq s,
$$

$$
\|\Psi_\mu(t, s)\| \leq K e^{\alpha_2(\mu)(t-s)} \text{ for all } \mu \in (\mu_-, \mu_+) \text{ and } t, s \in I \text{ with } t \leq s.
$$

The functions q and d_i satisfy one of the following conditions.

$$
\begin{aligned}
0 &< \liminf_{\mu \to \mu_0} \inf_{t \in I} q(t, \mu) \leq \limsup_{\mu \to \mu_0} \sup_{t \in I} q(t, \mu) < \infty, \\
0 &< \liminf_{\mu \to \mu_0} \inf_{i \in \mathbb{A}} d_i(\mu) \leq \limsup_{\mu \to \mu_0} \sup_{i \in \mathbb{A}} d_i(\mu) < \infty
\end{aligned}
\tag{5.1.20}
$$

or

$$
\begin{aligned}
-\infty &< \liminf_{\mu \to \mu_0} \inf_{t \in I} q(t, \mu) \leq \limsup_{\mu \to \mu_0} \sup_{t \in I} q(t, \mu) < 0, \\
-\infty &< \liminf_{\mu \to \mu_0} \inf_{i \in \mathbb{A}} d_i(\mu) \leq \limsup_{\mu \to \mu_0} \sup_{i \in \mathbb{A}} d_i(\mu) < 0.
\end{aligned}
\tag{5.1.21}
$$

The functions r and r_i satisfy the following conditions.

$$
\begin{aligned}
\lim_{x \to 0} \sup_{\mu \in (\mu_0 - x^2, \mu_0 + x^2)} \sup_{t \in I} \frac{|r(t, x, \mu)|}{|x|^3} &= 0, \\
\lim_{x \to 0} \sup_{\mu \in (\mu_0 - x^2, \mu_0 + x^2)} \sup_{i \in \mathbb{A}} \frac{|e_i(x, \mu)|}{|x|^3} &= 0
\end{aligned}
\tag{5.1.22}
$$

and there exists sufficiently small $l > 0$ such that

$$|r(t, x, \mu)| < l|x|, \quad |e_i(x, \mu)| < l|x|, \tag{5.1.23}$$

for all $\mu \in (\mu_-, \mu_+)$, $t \in I$, $i \in \mathbb{A}$ and $x \in (x_-, x_+)$.

Theorem 5.1.4 *Assume that above conditions hold for the system (5.1.18). Then there exists $\widehat{\mu}_- < 0 < \widehat{\mu}_+$ such that*

- *If the functions α_1 and α_2 are monotone increasing, the origin is all-time attractive for $\mu \in (\widehat{\mu}_-, \mu_0)$ and all-time repulsive for $\mu \in (\mu_0, \widehat{\mu}_+)$. The following relations hold true.*

$$\lim_{\mu \to \mu_0^-} \mathscr{A}_0^\mu = 0 \quad and \quad \lim_{\mu \to \mu_0^+} \mathscr{R}_0^\mu = 0;$$

- *If the functions α_1 and α_2 are monotone decreasing, the origin is all-time repulsive for $\mu \in (\widehat{\mu}_-, \mu_0)$ and all-time attractive for $\mu \in (\mu_0, \widehat{\mu}_+)$. The following relations hold true.*

$$\lim_{\mu \to \mu_0^+} \mathscr{A}_0^\mu = 0 \quad and \quad \lim_{\mu \to \mu_0^-} \mathscr{R}_0^\mu = 0;$$

Hence, in both of the above cases the system (5.1.18) possesses an all-time bifurcation.

The proof of the theorem is similar to that of Theorem 5.1.3.

5.2 Finite-Time Nonautonomous Bifurcations

The purpose of this section is to investigate nonautonomous bifurcation in impulsive differential equations in the finite-time interval. The impulsive finite-time analogues of transcritical and pitchfork bifurcation are provided. An illustrative example is given with numerical simulations which supports theoretical results.

5.2.1 Introduction and Preliminaries

In considering application problems which arise in the real world such as ocean or atmosphere dynamics [187], transport problems in fluid or any model in biological applications [58, 206], one may come across with finite dynamics. There are several reasons why bounded set dynamics is of a great importance. The first and the most simple reason is that data obtained from measurements and observations are often given in a compact time interval. Another reason may be the interest in transient behavior of solutions regardless of time interval in which differential equation is defined. Therefore, there is increasing interest in the behavior of the system on bounded time interval coming from application point of view.

The main novelty of this section is to provide suitable and efficient concepts of finite-time bifurcation in the context of nonautonomous differential equations with impulses.

We denote by $\mathbb{R}$ the set of all real numbers, $\mathbb{Z}$ the set of integers. In this section, we introduce concepts of attractive and repulsive solutions, which are used to analyze asymptotic behavior of impulsive nonautonomous systems. This section is concerned with systems of the type

$$\begin{aligned} x' &= f(t, x), \\ \Delta x|_{t=\theta_i} &= J_i(x), \end{aligned} \tag{5.2.24}$$

where $\Delta x|_{t=\theta_i} := x(\theta_i+) - x(\theta_i)$, $x(\theta_i+) = \lim_{t \to \theta_i^+} x(t)$. The system (5.2.24) is defined on the set $\Omega = I \times \mathbb{A} \times G$ where $G \subseteq \mathbb{R}^n$, $I \subset \mathbb{R}$ is a finite compact time interval which contains only a finite number of impulse points θ_i with the set of indexes $\mathbb{A}$. Let $\phi(t, t_0, x_0)$ be general solution of (5.2.24) which is unique and non-continuable.

Asymptotic properties of continuous dynamics and dynamics with discontinuous are the same. In what follows we use definitions of attractivity and repulsivity without any changes form [204, 205].

Definition 5.2.1 Let $t_0 \in I$ and $T > 0$ is such that $t_0 + T \in I$.

- A compact and invariant nonautonomous set $\mathcal{G}$ is called $(t_0, T) - attractor$ if

$$\limsup_{\varepsilon \to 0^+} \frac{1}{\varepsilon} d(\phi(t_0 + T, t_0, B_\varepsilon(\mathcal{G}(t_0))), \mathcal{G}(t_0 + T)) < 1;$$

- A solution $\psi : [t_0, t_0 + T] \to \mathbb{R}^n$ of (5.2.24) is called $(t_0, T) - attractive$ if $graph\psi$ is a $(t_0, T) - attractor$.
- A compact and invariant nonautonomous set $\mathcal{H}$ is called $(t_0, T) - repeller$ if

$$\limsup_{\varepsilon \to 0^+} \frac{1}{\varepsilon} d(\phi(t_0 + T, t_0, B_\varepsilon(\mathcal{H}(t_0 + T))), \mathcal{H}(t_0)) < 1.$$

- A solution $\psi : [t_0, t_0 + T] \to \mathbb{R}^n$ of (5.2.24) is called $(t_0, T) - repulsive$ if $graph\psi$ is a $(t_0, T) - repeller$.

The $(t_0, T) - attractor$ and $(t_0, T) - repeller$ satisfy the duality principle, i.e., under time reversal their roles are changed.

Example 5.2.1 Let $I := [t_0, t_0 + T]$ be an interval containing a finite number of impulse points θ_i such that $t_0 \leq \theta_1 < \theta_2 < \ldots < \theta_m \leq t_0 + T$ for some $t_0 \in \mathbb{R}, T > 0$ and $m \in \mathbb{N}$. Consider the system

$$\begin{aligned} x' &= a(t)x, \\ \Delta x|_{t=\theta_i} &= b_i x, \end{aligned} \tag{5.2.25}$$

with piecewise continuous function $a : I \to \mathbb{R}$ and there exist constants $b, B \in \mathbb{R}$ such that $-1 < b \le b_i \le B$. Let $\Psi : I \times I \to \mathbb{R}^n$ be the transition matrix of the system (5.2.24).

If $t_0 < \theta_1$, then $a(t)$ is continuous on $[t_0, \theta_1]$ since $a(t) \in PC(\mathbb{R}, \theta)$. So, we have that $\Psi(\theta_1, t_0) = e^{\int_{t_0}^{\theta_1} a(s)ds}$. At $t = \theta_1$, solution makes a jump and we have that $x(\theta_1+) = (1 + b_1)x(\theta_1)$. Next, $a(t)$ is continuous on $(\theta_1, \theta_2]$, implying that $\Psi(\theta_2, \theta_1) = e^{\int_{\theta_1}^{\theta_2} a(s)ds}(1 + b_1)$. Proceeding in this way one can show that

$$\Psi(t_0 + T, t_0) = \Psi(\theta_1, t_0)\Psi(\theta_2, \theta_1) \cdots \Psi(t_0 + T, \theta_m) = e^{\int_{t_0}^{t_0+T} a(s)ds} \prod_{i=1}^{m}(1 + b_i)$$

since $1 + b_i$ is nonsingular matrix and commutes with any other matrix because $1 + b_i > 0$.

If $t_0 = \theta_1$, then the solution starts with a jump and we have that $x(\theta_1+) = (1 + b_1)x(\theta_1)$. Next, $a(t)$ is continuous on $(\theta_1, \theta_2]$, implying that $\Psi(\theta_2, \theta_1) = e^{\int_{\theta_1}^{\theta_2} a(s)ds}(1 + b_1)$. Discussing in the way, one can show that

$$\Psi(t_0 + T, t_0) = \Psi(\theta_2, \theta_1) \cdots \Psi(t_0 + T, \theta_m) = e^{\int_{t_0}^{t_0+T} a(s)ds} \prod_{i=1}^{m}(1 + b_i).$$

We want to point out that the basics of linear impulsive systems are fruitfully discussed in the books [1, 156, 216]. As a result, we have that

$$\Psi(t_0 + T, t_0) = e^{\int_{t_0}^{t_0+T} a(s)ds} \prod_{i=1}^{m}(1 + b_i) \le e^{\int_{t_0}^{t_0+T} a(s)ds} \prod_{i=1}^{m}(1 + B)$$
$$= e^{\int_{t_0}^{t_0+T} a(s)ds + m\ln(1+B)}.$$

Therefore, any invariant and compact nonautonomous set is a $(t_0, T) - attractor$ if

$$\int_{t_0}^{t_0+T} a(s)ds + m\ln(1 + B) < 0.$$

By the same way one can say that any invariant and compact nonautonomous set is a $(t_0, T) - repeller$ if $\int_{t_0}^{t_0+T} a(s)ds + m\ln(1 + b) > 0$.

Definition 5.2.2 The radius of $(t_0, T) - attraction$ of a $(t_0, T) - attractor A$ is defined by
$\mathscr{A}_{\mathscr{G}}^{(t_0,T)} := \sup\{\varepsilon > 0 : d(\phi(t_0+T, t_0, B_{\widehat{\varepsilon}}(\mathscr{G}(t_0))), \mathscr{G}(t_0+T)) < \widehat{\varepsilon} \text{ for all } \widehat{\varepsilon} \in (0, \varepsilon)\}$,
and the radius of $(t_0, T) - repulsion$ of a $(t_0, T) - repeller R$ is defined by $\mathscr{R}_{\mathscr{H}}^{(t_0,T)} := \sup\{\varepsilon > 0 : d(\phi(t_0 + T, t_0, B_{\widehat{\varepsilon}}(\mathscr{H}(t_0 + T))), \mathscr{H}(t_0)) < \widehat{\varepsilon}$
for all $\widehat{\varepsilon} \in (0, \varepsilon)\}$.

In Example 5.2.1, every invariant and compact set $S \subset [t_0, t_0 + T] \times \mathbb{R}$ of the system (5.2.25) is

- $(t_0, T) - attractor$ with $\mathscr{A}_S^{(t_0,T)} = \infty$ if $\int_{t_0}^{t_0+T} a(s)ds + m\ln(1 + B) < 0$,
- $(t_0, T) - repeller$ with $\mathscr{R}_S^{(t_0,T)} = \infty$ if $\int_{t_0}^{t_0+T} a(s)ds + m\ln(1 + b) > 0$.

Definition 5.2.3 We consider the impulsive system (5.2.24), which depends on a parameter μ. The system (5.2.24) possesses a supercritical $(t_0, T) - bifurcation$ at μ_0, $\mu_0 \in (\mu_-, \mu_+)$, if there exists a $\widehat{\mu} > \mu_0$ and a piecewise continuous function $\psi : [t_0, t_0 + T] \times (\mu_0, \widehat{\mu}) \to \mathbb{R}^n$ such that one of the following alternatives hold.

- $\psi(\cdot, \mu)$ is a $(t_0, T) - attractive$ solution of the system (5.2.24) for all $\mu \in (\mu_0, \widehat{\mu})$, and

$$\lim_{\mu \to \mu_0^+} \mathscr{A}_{\psi(\cdot,\mu)}^{(t_0,T)} = 0.$$

- $\psi(\cdot, \mu)$ is a $(t_0, T) - repulsive$ solution of the system (5.2.24) for all $\mu \in (\mu_0, \widehat{\mu})$, and

$$\lim_{\mu \to \mu_0^+} \mathscr{R}_{\psi(\cdot,\mu)}^{(t_0,T)} = 0.$$

If in the above definition the limit $\mu \to \mu_0^+$ is replaced with $\mu \to \mu_0^-$, then we have subcritical $(t_0, T) - bifurcation$.

5.2.2 Attractivity and Repulsivity in a Linear Nonhomogeneous Impulsive System

In this section, we study linear nonhomogeneous impulsive systems in finite time with definitions provided in the previous section. We show that these definitions remain as a fruitful concept in the stability analysis of solutions of nonlinear systems with fixed moments of impulses. Let us consider the impulsive system in a compact interval $I := [t_0, t_0 + T]$ with m impulse points θ_i for some $t_0 \in \mathbb{R}, T > 0$ and $m \in \mathbb{N}$,

$$\begin{aligned} x' &= A(t)x + F(t, x), \\ \Delta x|_{t=\theta_i} &= B_i x + J_i(x), \end{aligned} \tag{5.2.26}$$

where $A \in PC(I, \theta)$, matrices B_i satisfy $\det(B_i + I) \neq 0$, $F : I \times G \to \mathbb{R}^n$ and $J : \mathbb{A} \times G \to \mathbb{R}^n$. Denote $\phi(t, t_0, x_0)$ as the solution of (5.2.26) and $\Psi : I \times I \to \mathbb{R}^{n \times n}$ as the transition matrix of the following linear system

$$\begin{aligned} x' &= A(t)x, \\ \Delta x|_{t=\theta_i} &= B_i x. \end{aligned} \tag{5.2.27}$$

Define $M_+ := \sup \{||\Psi(t,s)|| : t_0 \leq s \leq t \leq t_0 + T\}$ and
$M_- := \sup \{||\Psi(t,s)|| : t_0 \leq t \leq s \leq t_0 + T\}$. Assume that the following conditions hold for the system (5.2.26):

(C1) $||\Psi(t_0 + T, t_0)|| < 1$;
(C2) The functions $F(t,x)$ and $J_i(x)$ are Lipschitzian, i.e., there exists positive number l such that $||F(t,x)|| \leq l||x||$, $||J_i(x)|| \leq l||x||$ for all $t \in I, i \in \mathbb{A}$ and $||x|| < h, h > 0$.

Then, one has the following theorem.

Theorem 5.2.1 *The origin is (t_0, T)-attractive for sufficiently small values of l, that is,*

$$||\phi(t_0 + T, t_0, x_0)|| \leq \delta e^{M_+ Tl + m\ln(1+M_+l) + \ln||\Psi(t_0+T,t_0)||}.$$

Now consider the following condition

(C3) $||\Psi(t_0, t_0 + T)|| < 1$.

Theorem 5.2.2 *Assume that conditions (C2) and (C3) are true for the system (5.2.26), then the origin is (t_0, T)-repulsive for sufficiently small values of l, i.e.,*

$$||\phi(t_0, t_0 + T, x_0)|| \leq \delta e^{M_- Tl + m\ln(1+M_-l) + \ln||\Psi(t_0,t_0+T)||}.$$

Proof An equivalent integral equation of the system (5.2.26) can be written as [1, 216]:

$$\phi(t, t_0, x_0) = \Psi(t, t_0)x_0 + \int_{t_0}^t \Psi(t, s)F(s, \phi(s, t_0, x_0))ds$$
$$+ \sum_{t_0 \leq \theta_i < t} \Psi(t, \theta_i)I_i(\phi(\theta_i, t_0, x_0))$$

for all $t \in I$. Therefore, we get

$$||\phi(t, t_0, x_0)|| \leq ||\Psi(t, t_0)||||x_0|| + M_+ l \int_{t_0}^t ||\phi(s, t_0, x_0)||ds$$
$$+ M_+ l \sum_{t_0 \leq \theta_i < t} ||\phi(\theta_i, t_0, x_0)||$$

for all $t \in I$ is fulfilled. Hence, by Gronwall–Bellman lemma for piecewise continuous functions [1, 216] it follows that

$$||\phi(t_0 + T, t_0, x_0)|| \leq ||\Psi(t_0 + T, t_0)||e^{M_+ lT}(1 + M_+l)^{i[t_0,t_0+T)}||x_0||$$
$$\leq ||x_0||e^{\ln||\Psi(t_0+T,t_0)|| + M_+ lT + m\ln(1+M_+l)}$$

where $i[t_0, t_0 + T)$ is the number of elements of the sequence θ_i in the interval $[t_0, t_0 + T)$. Since in this section $i[t_0, t_0 + T) = m$, one can see that the required result follows by choosing $\delta = Kh$ for l small enough that $\ln \|\Psi(t_0 + T, t_0)\| + M_+ lT + m \ln(1 + M_+ l) < 0$. We skip the prove of Theorem 5.2.2 since it can be proven analogously. $\square$

5.2.3 Bifurcation Analysis

In this section, we state and prove finite-time nonautonomous transcritical and pitchfork bifurcation results for impulsive systems. In what follows, the auxiliary theorems obtained in the previous section for higher dimensions will be used in the scalar case.

5.2.3.1 The Transcritical Bifurcation

In this subsection, we study impulsive analogue of the nonautonomous transcritical bifurcation in finite time. Let $x_- < 0 < x_+$ and $\mu_- < \mu_+$ be real numbers and let $I := [t_0, t_0 + T]$ with m impulse points θ_i. Consider the system

$$\begin{aligned} x' &= p(t, \mu)x + q(t, \mu)x^2 + r(t, x, \mu), \\ \Delta x|_{t=\theta_i} &= a_i(\mu)x + b_i(\mu)x^2 + c_i(x, \mu), \end{aligned} \tag{5.2.28}$$

with piecewise continuous functions $p : I \times (\mu_-, \mu_+) \to \mathbb{R}$, $q : I \times (\mu_-, \mu_+) \to \mathbb{R}$ and $r : I \times (x_-, x_+) \times (\mu_-, \mu_+) \to \mathbb{R}$ satisfying $r(t, 0, \mu) = 0$. $a : \mathbb{A} \times (\mu_-, \mu_+) \to \mathbb{R}$, $b : \mathbb{A} \times (\mu_-, \mu_+) \to \mathbb{R}$ and $c : \mathbb{A} \times (x_-, x_+) \times (\mu_-, \mu_+) \to \mathbb{R}$ with $a_i(\mu) \neq -1$ and $c_i(0, \mu) = 0$. Let $\Psi_\mu(t, s)$ be the fundamental matrix of the associated homogeneous part of the system

$$\begin{aligned} x' &= p(t, \mu)x, \\ \Delta x|_{t=\theta_i} &= a_i(\mu)x. \end{aligned} \tag{5.2.29}$$

Assume that there exists $\mu_0 \in (\mu_-, \mu_+)$ such that the following conditions hold.

(T1) $\Psi_\mu(t_0 + T, t_0) < 1$ for all $\mu \in (\mu_-, \mu_0)$ and $\Psi_\mu(t_0 + T, t_0) > 1$ for all $\mu \in (\mu_0, \mu_+)$;
   or
(T1*) $\Psi_\mu(t_0 + T, t_0) > 1$ for all $\mu \in (\mu_-, \mu_0)$ and $\Psi_\mu(t_0 + T, t_0) < 1$ for all $\mu \in (\mu_0, \mu_+)$.

The functions q and b_i satisfy one of the following conditions.

(T2) $\liminf_{\mu \to \mu_0} \inf_{t \in I} q(t, \mu) > 0$ and $\liminf_{\mu \to \mu_0} \inf_{i \in \mathbb{A}} b_i(\mu) > 0$;
   or
(T2*) $\limsup_{\mu \to \mu_0} \sup_{t \in I} q(t, \mu) < 0$ and $\limsup_{\mu \to \mu_0} \sup_{i \in \mathbb{A}} b_i(\mu) < 0$.

The functions r and c_i satisfy the following conditions.

(T3) $\lim_{x \to 0} \sup_{\mu \in (\mu_0 - |x|, \mu_0 + |x|)} \sup_{t \in I} \frac{|r(t, x, \mu)|}{|x|^2} = 0$;

(T4) $\lim_{x \to 0} \sup_{\mu \in (\mu_0 - |x|, \mu_0 + |x|)} \sup_{i \in \mathbb{A}} \frac{|c_i(x, \mu)|}{|x|^2} = 0$;

(T5) There exists sufficiently small $l > 0$ such that $|r(t, x, \mu)| < l|x|$, $|c_i(x, \mu)| < l|x|$ for all $\mu \in (\mu_-, \mu_+)$, $t \in I$, $i \in \mathbb{A}$ and $x \in (x_-, x_+)$.

Theorem 5.2.3 *Assume that above conditions hold for the system (5.2.28). Then there exists $\widehat{\mu}_- < 0 < \widehat{\mu}_+$ such that if (T1) is satisfied, then the origin is $(t_0, T) -$ attractive for $\mu \in (\widehat{\mu}_-, \mu_0)$ and $(t_0, T) -$ repulsive for $\mu \in (\mu_0, \widehat{\mu}_+)$. The following relations hold true.*

$$\lim_{\mu \to \mu_0^-} \mathscr{A}_0^\mu = 0 \quad \text{and} \quad \lim_{\mu \to \mu_0^+} \mathscr{R}_0^\mu = 0. \tag{5.2.30}$$

In case (T1*) is satisfied, the origin is $(t_0, T) -$ repulsive for $\mu \in (\widehat{\mu}_-, \mu_0)$ and $(t_0, T) -$ attractive for $\mu \in (\mu_0, \widehat{\mu}_+)$. The following relations hold true.

$$\lim_{\mu \to \mu_0^+} \mathscr{A}_0^\mu = 0 \quad \text{and} \quad \lim_{\mu \to \mu_0^-} \mathscr{R}_0^\mu = 0. \tag{5.2.31}$$

Hence, in both of the above cases the system (5.2.28) possesses a $(t_0, T) -$ bifurcation.

Proof We give the first part of the proof since second part can be proven in the similar manner. That is, (T1) is assumed. Let ϕ_μ be the general solution of the system (5.2.28). We may consider the case (T2). Choose $\widehat{\mu}_- < \mu_0 < \widehat{\mu}_+$ such that

$$0 < \inf_{\mu \in (\widehat{\mu}_-, \widehat{\mu}_+), t \in I} q(t, \mu) \quad \text{and} \quad 0 < \inf_{\mu \in (\widehat{\mu}_-, \widehat{\mu}_+), i \in \mathbb{A}} b_i(\mu). \tag{5.2.32}$$

By means of (T4) and (5.2.32), one can see that Theorems 5.2.1 and 5.2.2 can be applied. Thus, we get attractivity and repulsivity of the origin as it was required to show in the theorem. Set

$$K := \inf \left\{ \Psi_\mu(t, s) : t, s \in I, \mu \in (\widehat{\mu}_-, \mu_0) \right\} \in (0, 1).$$

In order to show relations (5.2.30) and (5.2.31) hold, we assume to the contrary that

$$\gamma = \limsup_{\mu \to \mu_0^-} \mathscr{A}_0^\mu > 0.$$

By means of (T3) and (5.2.32) one can show that there exists $\widetilde{\mu} \in (\widehat{\mu}_-, \mu_0)$, $x_0 \in (0, K\gamma)$ and $L > 0$ such that

$$q(t, \mu)x^2 + r(t, x, \mu) > J \quad \text{and} \quad b_i(\mu)x^2 + c_i(x, \mu) > J \tag{5.2.33}$$

for all $t \in I$, $i \in \mathbb{A}$, $\mu \in (\widetilde{\mu}, \mu_0)$ and $x_0 \in \left[Kx_0, \frac{x_0}{K} \right]$. Next, fix $\widehat{\mu} \in (\widetilde{\mu}, \mu_0)$ such that $\mathscr{A}_0^{\widehat{\mu}} > x_0$ and

$$\Psi_\mu(t_0 + T, t_0) \geq 1 - \frac{KJT}{x_0}. \tag{5.2.34}$$

Denote $\psi(t) = \phi_{\widehat{\mu}}(t, t_0, x_0)$. Since $\mathscr{A}_0^{\widehat{\mu}} > x_0$, we have

$$\psi(t_0 + T) < x_0. \tag{5.2.35}$$

Moreover, from the definition of K and by means of (5.2.33), we get

$$\psi(t_0 + T) \geq Kx_0 \quad \text{for all} \ t \in [0, T]. \tag{5.2.36}$$

There are two cases to be considered.

Case 1. There exists a $\tau \in (0, T]$ such that

$$\psi(t_0 + \tau) = \frac{x_0}{K}.$$

We take maximal τ which satisfies this relation. By means of (5.2.35), one can see that $\psi(t_0 + T) \leq \frac{x_0}{K}$ for all $t \in [\tau, T]$. Next, we consider the integral equation of the system (5.2.28) at $t_0 + T$ for fixed $\widehat{\mu}$ which starts at point $t = t_0 + \tau$.

$$\begin{aligned}
\psi(t_0 + T) &= \Psi_{\widehat{\mu}}(t_0 + T, t_0 + \tau)\frac{x_0}{K} \\
&\quad + \int_{t_0+\tau}^{t_0+T} \Psi_{\widehat{\mu}}(t_0 + T, s)\left(q(s, \widehat{\mu})(\psi(s))^2 + r(s, \psi(s), \widehat{\mu})\right) ds \\
&\quad + \sum_{t_0+\tau \leq \theta_i < t_0+T} \Psi_{\widehat{\mu}}(t_0 + T, \theta_i)\left(b_i(\widehat{\mu})(\psi(\theta_i))^2 + c_i(\psi(\theta_i), \widehat{\mu})\right) \\
&\geq x_0 + KJ(T - \tau) + \sum_{t_0+\tau \leq \theta_i < t_0+T} KJ \\
&> x_0.
\end{aligned}$$

This is contraction for (5.2.35).

Case 2. For all $t \in (0, T]$, we have

$$\psi(t_0 + \tau) < \frac{x_0}{K}.$$

Next, from the integral equation of the system (5.2.28) at $t = t_0 + T$ for fixed $\widehat{\mu}$ which starts at point t_0 we have

$$\begin{aligned}
\psi(t_0 + T) &= \Psi_{\widehat{\mu}}(t_0 + T, t_0 + \tau)x_0 \\
&\quad + \int_{t_0}^{t_0+T} \Psi_{\widehat{\mu}}(t_0 + T, s)\left(q(s, \widehat{\mu})(\psi(s))^2 + r(s, \psi(s), \widehat{\mu})\right) ds \\
&\quad + \sum_{t_0 \leq \theta_i < t_0+T} \Psi_{\widehat{\mu}}(t_0 + T, \theta_i)\left(b_i(\widehat{\mu})(\psi(\theta_i))^2 + c_i(\psi(\theta_i), \widehat{\mu})\right) \\
&\geq \left(1 - \frac{KJT}{x_0}\right)x_0 + KJT + KJm \\
&> x_0
\end{aligned}$$

where the last inequality follows by means of (5.2.33) and (5.2.34). We again arrive at contradiction for (5.2.35). Hence, we have that $\lim_{\mu \to \mu_0^-} \mathscr{A}_0^\mu = 0$. One can show following the similar route that $\lim_{\mu \to \mu_0^+} \mathscr{R}_0^\mu = 0$ and considering the condition (T2*). This finalizes the proof of the theorem. $\square$

5.2.3.2 The Pitchfork Bifurcation

In this subsection, we study impulsive analogue of the nonautonomous pitchfork bifurcation. Let $x_- < 0 < x_+$ and $\mu_- < \mu_+$ be real numbers and let $I := [t_0, t_0 + T]$ with m impulse points θ_i. Consider the system

$$x' = p(t, \mu)x + q(t, \mu)x^3 + r(t, x, \mu),$$
$$\Delta x|_{t=\theta_i} = a_i(\mu)x + b_i(\mu)x^3 + c_i(x, \mu), \qquad (5.2.37)$$

with piecewise continuous functions $p : I \times (\mu_-, \mu_+) \to \mathbb{R}, q : I \times (\mu_-, \mu_+) \to \mathbb{R}$ and $r : I \times (x_-, x_+) \times (\mu_-, \mu_+) \to \mathbb{R}$ satisfying $r(t, 0, \mu)=0$. $a : \mathbb{A} \times (\mu_-, \mu_+) \to \mathbb{R}, b : \mathbb{A} \times (\mu_-, \mu_+) \to \mathbb{R}$ and $c : \mathbb{A} \times (x_-, x_+) \times (\mu_-, \mu_+) \to \mathbb{R}$ with $a_i(\mu) \neq -1$ and $c_i(0, \mu) = 0$. Let $\Psi_\mu(t, s)$ be the fundamental matrix of the linear system

$$x' = p(t, \mu)x,$$
$$\Delta x|_{t=\theta_i} = a_i(\mu)x.$$

Assume that there exists $\mu_0 \in (\mu_-, \mu_+)$ such that following conditions hold.

(P1) $\Psi_\mu(t_0 + T, t_0) < 1$ for all $\mu \in (\mu_-, \mu_0)$ and $\Psi_\mu(t_0 + T, t_0) > 1$ for all $\mu \in (\mu_0, \mu_+)$;
   or
(P1*) $\Psi_\mu(t_0 + T, t_0) > 1$ for all $\mu \in (\mu_-, \mu_0)$ and $\Psi_\mu(t_0 + T, t_0) < 1$ for all $\mu \in (\mu_0, \mu_+)$.

The functions q and b_i satisfy one of the following conditions.

(P2) $\liminf_{\mu \to \mu_0} \inf_{t \in I} q(t, \mu) > 0$ and $\liminf_{\mu \to \mu_0} \inf_{i \in \mathbb{A}} b_i(\mu) > 0$;
   or
(P2*) $\limsup_{\mu \to \mu_0} \sup_{t \in I} q(t, \mu) < 0$ and $\limsup_{\mu \to \mu_0} \sup_{i \in \mathbb{A}} b_i(\mu) < 0$.

The functions r and c_i satisfy the following conditions.

(P3) $\lim_{x \to 0} \sup_{\mu \in (\mu_0 - x^2, \mu_0 + x^2)} \sup_{t \in I} \frac{|r(t,x,\mu)|}{|x|^3} = 0$;
(P4) $\lim_{x \to 0} \sup_{\mu \in (\mu_0 - x^2, \mu_0 + x^2)} \sup_{i \in \mathbb{A}} \frac{|c_i(x,\mu)|}{|x|^3} = 0$;
(P5)   There exists sufficiently small $l > 0$ such that $|r(t, x, \mu)| < l|x|, |c_i(x, \mu)| < l|x|$ for all $\mu \in (\mu_-, \mu_+), t \in I, i \in \mathbb{A}$ and $x \in (x_-, x_+)$.

Theorem 5.2.4 *Assume that above conditions hold for the system (5.2.37). Then there exists $\widehat{\mu}_- < 0 < \widehat{\mu}_+$ such that if the conditions (P1) and (P2) are satisfied, then the origin is $(t_0, T) - attractive$ for $\mu \in (\widehat{\mu}_-, \mu_0)$ and $(t_0, T) - repulsive$ for*

$\mu \in (\mu_0, \widehat{\mu}_+)$. *The following relation is satisfied.*

$$\lim_{\mu \to \mu_0^-} \mathscr{A}_0^\mu = 0.$$

If the conditions (P1) and (P2*) are satisfied, then the origin is (t_0, T) − repulsive for $\mu \in (\widehat{\mu}_-, \mu_0)$ and (t_0, T) − attractive for $\mu \in (\mu_0, \widehat{\mu}_+)$. The following relation is satisfied.

$$\lim_{\mu \to \mu_0^+} \mathscr{R}_0^\mu = 0.$$

If the conditions (P1*) and (P2) hold, the origin is (t_0, T) − attractive for $\mu \in (\widehat{\mu}_-, \mu_0)$ and (t_0, T) − repulsive for $\mu \in (\mu_0, \widehat{\mu}_+)$. The following relation is satisfied.

$$\lim_{\mu \to \mu_0^+} \mathscr{A}_0^\mu = 0.$$

In case the conditions (P1*) and (P2*) hold, the origin is (t_0, T) − repulsive for $\mu \in (\widehat{\mu}_-, \mu_0)$ and (t_0, T) − attractive for $\mu \in (\mu_0, \widehat{\mu}_+)$. The following relation is satisfied.

$$\lim_{\mu \to \mu_0^-} \mathscr{R}_0^\mu = 0.$$

Hence, in all of the above cases the system (5.2.37) *possesses a (t_0, T) − bifurcation*

The proof of the theorem is similar to that of Theorem 5.2.3.

5.2.4 An Example

In this section, we give an example illustrating our theoretical results by means of simulations. Consider the following system with $I := [0, 10]$ and impulse moments $\theta_i = i$, $1 \le i \le 9$.

$$x' = \left(6\mu + \tfrac{5}{2}\mu \sin(\tfrac{t^3}{4})\right) x - \left(4\mu + \tfrac{7}{2}\mu \sin(\tfrac{t^5}{3}) + 2\right) x^2 + \left(\mu + \tfrac{1}{2}\mu \cos^2(t^3)\right) x^3,$$
$$\Delta x|_{t=i} = (1.5i\mu + 5\mu) x - (2i\mu + 5\mu + 3) x^2 + i\mu x^3,$$

where we have taken $p(t, \mu) = 6\mu + 2.5\mu \sin(t^3/4)$, $q(t, \mu) = 4\mu + 3.5\mu \sin(t^5/3) + 2$, $r(t, x, \mu) = \left(\mu + 0.5\mu \cos^2(t^3)\right) x^3$ $a_i(\mu) = 1.5i\mu + 5\mu$, $b_i(\mu) = 2i\mu + 5\mu + 3$ and $d_i(x, \mu) = i\mu x^3$. One can verify that this system satisfies the conditions of Theorem 5.2.3. Simulation results support our theoretical discussion and reveal that all solutions starting in the neighborhood of the origin converge to the origin if $\mu < 0$, whereas for $\mu < 0$ all solutions starting in the neighborhood of the origin diverge from the origin.

Fig. 5.1 Asymptotic behavior of the solution for $\mu = -0.1$, where *blue* color represents the solution corresponding to $x_0 = 0.4$; *red* color represents the solution corresponding to $x_0 = 0.1$; and *green* color represents the solution corresponding to $x_0 = -0.2$

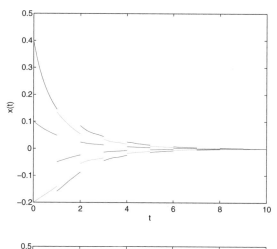

Fig. 5.2 Asymptotic behavior of the solution for $\mu = -0.05$, where *blue* color represents the solution corresponding to $x_0 = 0.4$; *red* color represents the solution corresponding to $x_0 = 0.1$; and *green* color represents the solution corresponding to $x_0 = -0.2$

Fig. 5.3 Asymptotic behavior of the solution for $\mu = 0.05$, where *blue* color represents the solution corresponding to $x_0 = 0.4$; *red* color represents the solution corresponding to $x_0 = 0.1$; and *green* color represents the solution corresponding to $x_0 = 0.2$

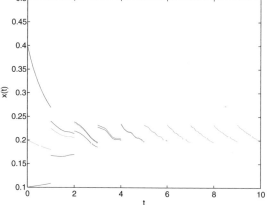

Fig. 5.4 Asymptotic
behavior of the solution for
$\mu = 0.1$, where *blue* color
represents the solution
corresponding to $x_0 = 0.4$;
red color represents the
solution corresponding to
$x_0 = 0.1$; and *green* color
represents the solution
corresponding to $x_0 = 0.2$

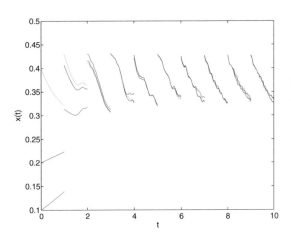

From the simulation results, it is seen that $\lim_{\mu \to 0^-} \mathscr{A}_0^{\mu} = 0$ since solutions in Fig. 5.2 converge more rapidly to the origin than those in Fig. 5.1; and $\lim_{\mu \to 0^+} \mathscr{R}_0^{\mu} = 0$ since solutions in Fig. 5.4 diverge more rapidly from the origin than those in Fig. 5.3. Thus, the origin is $(0, 10) - attractive$ for $\mu \in (-0.1, 0)$ and $(0, 10) - repulsive$ for $\mu \in (0, 0.1)$. We conclude that this example possesses $(0, 10)-$ transcritical bifurcation.

5.3   Notes

The results of this chapter were published in [33, 34]. There are qualitative papers devoted to nonautonomous bifurcation theory in continuous dynamical systems studied in the last twenty years [138, 139, 152, 154, 204, 205]. In this chapter, it is the first time nonautonomous transcritical and pitchfork bifurcations are studied for the most general impulsive systems with fixed time. Bifurcation patterns are obtained through loss of attractivity and gain of repulsivity. Furthermore, we discussed finite-time bifurcation scenarios in the scalar discontinuous systems which are applicable for the real-world problems. We show that attractive and repulsive solutions play a great role in the theory of bifurcations. Numerical simulations suggest us that, as in Chap. 4, asymptotic behavior of the bounded solutions should also be examined for these types of equations. Finally, we note that this theory can be extended for higher-dimensional systems.

Chapter 6
Nonautonomous Bifurcations in Bernoulli Differential Equations with Piecewise Constant Argument of Generalized Type

In this chapter, we study existence of the bounded solutions and asymptotic behavior of the Bernoulli equations with piecewise constant argument. Nonautonomous pitchfork and transcritical bifurcation scenarios are investigated. An example with numerical simulation is given to illustrate our results.

6.1 Introduction and Preliminaries

The Bernoulli equations are important class of nonlinear systems in the classical theory of differential equation. Even though these equations have nonlinearities in continuous case, it is possible to obtain the exact solution by certain transformations. In Chap. 4, we investigated the Bernoulli equations with impulses. To the best of our knowledge, there is no study which deals with the Bernoulli equations with piecewise constant argument.

Let $\mathbb{Z}$, $\mathbb{N}$, and $\mathbb{R}$ be the set of all integers, natural, and real numbers, respectively. Fix two real-valued sequences θ_i, ζ_i, $i \in \mathbb{Z}$, such that $\theta_i < \theta_{i+1}$, $\theta_i \le \zeta_i \le \theta_{i+1}$ for all $i \in \mathbb{Z}$, $|\theta_i| \to \infty$ as $|i| \to \infty$. The main subject under investigation in this chapter is the following Bernoulli equation with piecewise constant argument of generalized type.

$$x'(t) = p(t)x(t) - q(t)x^n(t)x^n(\gamma(t)), \qquad (6.1.1)$$

where $x \in \mathbb{R}$, $t \in \mathbb{R}$, $\gamma(t) = \zeta_i$, if $t \in [\theta_i, \theta_{i+1})$, $i \in \mathbb{Z}$, the functions $p, q : \mathbb{R} \to \mathbb{R}$ are continuous. Let $\phi(t, \gamma(t), t_0, x_0)$ be a solution of (6.1.1). In this chapter, we treat only scalar differential equations such that $\phi(t, \gamma(t), t_0, x_0)$ is continuable on $\mathbb{R}$. Solutions are unique both forward and backward in time.

© Springer Nature Singapore Pte Ltd. and Higher Education Press 2017
M. Akhmet and A. Kashkynbayev, *Bifurcation in Autonomous and Nonautonomous Differential Equations with Discontinuities*, Nonlinear Physical Science, DOI 10.1007/978-981-10-3180-9_6

In this chapter, we consider differential equation with both retarded and advanced piecewise constant argument of generalized type. That is, the argument function, $\gamma(t)$, is of a mixed type. The nonlinear term of the Eq. (6.1.1) is taken so that after substitution $y(t) + y(\gamma(t)) = 2x^{1-n}(t)x^{-n}(\gamma(t))$, the system (6.1.1) is converted to a linear nonhomogeneous system. The main novelty of this chapter is that the Bernoulli equation with piecewise constant argument (6.1.1) is considered for the first time in the literature. The remaining part of this chapter is organized as follows: In Sect. 6.2, we study bounded solutions of the Bernoulli equation (6.1.1). In Sects. 6.3 and 6.4, the pitchfork and the transcritical bifurcations are investigated, respectively, along with asymptotic properties of the bounded solutions.

In order to study nonautonomous bifurcation with piecewise constant argument, we should define corresponding concepts of stability. In this chapter, we use Hausdorff semi-distance between sets X and Y as $d(X, Y) = \sup\limits_{x \in X} \inf\limits_{y \in Y} d(x, y)$.

6.1.1 Attraction and Stability

Asymptotic properties of continuous dynamics and hybrid dynamics are the same. Therefore, we shall use notion of pullback attracting sets without any change from [61, 75, 81, 85, 96–98, 140, 142, 153, 155, 204, 227].

Definition 6.1.1 ([140]) An invariant set $\mathfrak{A}(t)$ is called pullback attracting if for every $t \in \mathbb{R}$

$$\lim_{t_0 \to -\infty} d(\phi(t, t_0, x_0), \mathfrak{A}(t)) = 0.$$

Having given meanings of pullback attraction, one needs to characterize related ideas of stability, instability, and asymptotic stability in order to investigate asymptotic analysis in the pullback sense. Next, we start with defining stability in the pullback sense.

Definition 6.1.2 ([152]) An invariant set $\mathfrak{A}(t)$ is pullback stable if for every $t \in \mathbb{R}$ and $\varepsilon > 0$ there exists a $\delta(t) > 0$ such that for any $t_0 < t$, $x_0 \in N(\mathfrak{A}(t_0), \delta(t))$ implies that $\phi(t, t_0, x_0) \in N(\mathfrak{A}(t), \varepsilon)$.

An invariant set $\mathfrak{A}(t)$ is said to be pullback asymptotically stable if it is pullback stable and pullback attracting. As we are busy with scalar impulsive systems, one can verify that pullback attraction implies pullback stability for a bounded trajectory. Next, we state the following lemma which will be useful in what follows.

Lemma 6.1.1 *Let $y(t)$ be a locally pullback attracting complete trajectory of a scalar impulsive system. Then, $y(t)$ is also pullback stable.*

The proof of this lemma, given by Langa et al. in [154], for continuous case is the same for impulsive systems. Thus, the last lemma allows us to concentrate on only pullback attraction properties of a complete trajectory instead of carrying out pullback stability.

As one would expect, pullback instability is characterized through the converse of pullback stability. That is, an invariant set $\mathfrak{A}(t)$ is called pullback unstable if it is not pullback stable, i.e., if there exists a $t \in \mathbb{R}$ and $\varepsilon > 0$ such that for each $\delta > 0$, there exists a $t_0 < t$ and $x_0 \in N(\mathfrak{A}(t_0), \delta)$ such that $d(\phi(t, t_0, x_0), \mathfrak{A}(t)) > \varepsilon$. However, the notion of unstable set, which Crauel defined for the random dynamical systems, seems to be more natural instrument in discontinuous dynamics point of view.

Definition 6.1.3 ([96]) The unstable set, $U_{\mathfrak{A}(t)}$, of an invariant set $\mathfrak{A}(t)$ is defined as

$$U_{\mathfrak{A}(t)} = \{u : \lim_{t \to -\infty} d(\phi(t, t_0, u), \mathfrak{A}(t)) = 0\}.$$

We say that $\mathfrak{A}(t)$ is asymptotically unstable if the relation $U_{\mathfrak{A}(t)} \neq \mathfrak{A}(t)$ is fulfilled for some t.

If $\mathfrak{A}(t)$ is invariant, then one can see that $\mathfrak{A}(t) \subset U_{\mathfrak{A}(t)}$ is satisfied. Thus, from the last definition we have that $\mathfrak{A}(t)$ is strict subset of $U_{\mathfrak{A}(t)}$. In the sequel, we need the following result.

Proposition 6.1.1 ([152]) *If $\mathfrak{A}(t)$ is asymptotically unstable then it is also locally pullback unstable and cannot be locally pullback attracting.*

This result proven by Langa et al. in [152] is valid for impulsive systems. Thus, we omit the proof and refer to [152]. Note that the idea of the asymptotic instability is a definition of time-reversed forward attraction. Alternatively, it is conceivable to define instability as a time-reversed version of pullback attraction.

Definition 6.1.4 ([154]) An invariant set $\mathfrak{A}(t)$ is pullback repelling if it is pullback attracting for time-reversed system, i.e., if for every $t \in \mathbb{R}$ and every $x_0 \in \mathbb{R}^n$,

$$\lim_{t_0 \to \infty} d(\phi(t, t_0, x_0), \mathfrak{A}(t)) = 0.$$

6.2 Bounded Solutions

In this section, we study existence of a bounded solution of (6.1.1). It is easy to see that $x = 0$ is a bounded solution of (6.1.1). We are interested in nonzero bounded solutions. For this purpose, we shall need the following conditions.

(C1) There exist positive numbers m and M such that $0 < m \leq q(t) \leq M$ for all $t \in \mathbb{R}$;

(C2) There exist positive numbers $\underline{\theta}$ and $\overline{\theta}$ such that $\underline{\theta} \leq \theta_{i+1} - \theta_i \leq \overline{\theta}$.

By means of the substitution $y(t) + y(\gamma(t)) = 2x^{1-n}(t)x^{-n}(\gamma(t))$, the system (6.1.1) is reduced to the following nonhomogeneous linear system

$$y'(t) = (1 - n)p(t)y(t) + (1 - n)p(t)y(\gamma(t)) + 2(n - 1)q(t). \quad (6.2.2)$$

One can see that $\Psi(t, s) = e^{\int_s^t (1-n)p(u)du}$ is the fundamental solution of the following linear system

$$z'(t) = (1 - n)p(t)z(t).$$

In what follows, we introduce a function $R_i(t)$, $i \in \mathbb{Z}$, [2, 16],

$$R_i(t) = \Psi(t, \zeta_i) + \int_{\zeta_i}^t \Psi(t, s)(1 - n)p(s)ds = 2e^{\int_{\zeta_i}^t (1-n)p(u)du} - 1.$$

We shall need the following modified regularity condition.

(C3)   For every fixed $i \in \mathbb{Z}$, $R_i(t) > 0$, $\forall t \in [\theta_i, \theta_{i+1}]$.

Let $Y(t, s)$ be the fundamental matrix of the following linear system

$$y'(t) = (1 - n)p(t)y(t) + (1 - n)p(t)y(\gamma(t)). \tag{6.2.3}$$

Assume that $\theta_i < t_0 < \zeta_i$ for a fixed $i \in \mathbb{Z}$. One can confirm that [2, 7, 16],

$$Y(t, t_0) = R_j(t)\left(\prod_{k=j}^{i+1} R_k^{-1}(\theta_k)R_{k-1}(\theta_k)\right) R_i^{-1}(t_0),$$

for $t \in [\theta_j, \theta_{j+1}]$ and arbitrary $j > i$.

Denote $\alpha = \limsup\limits_{t-s \to \infty} \dfrac{\ln \|Y(t, s)\|}{t - s}$. One can guarantee that there exist two positive numbers k and K such that

$$ke^{\alpha(t-s)} \le \|Y(t, s)\| \le Ke^{\alpha(t-s)}, \quad s \le t. \tag{6.2.4}$$

In what follows, it will be useful to define the following piecewise continuous matrix.

$$\Sigma(t, s) = \begin{cases} Y(\theta_i, t_0)\Psi(t_0, s), & \text{if } t \in [t_0, \hat{\zeta}_i], \\ Y(t, \theta_{k+1})\Psi(\theta_{k+1}, s), & \text{if } t \in [\zeta_k, \zeta_{k+1}], \\ \Psi(t, s), & \text{if } t \in [\hat{\zeta}_j, t], \end{cases} \tag{6.2.5}$$

where $t_0 \in [\theta_i, \theta_{i+1}]$, $t \in [\theta_j, \theta_{j+1}]$, $i < j$, $[\hat{a, b}] = [a, b]$ if $a \le b$, and equal to $[b, a]$, otherwise for $a, b \in \mathbb{R}$.

Lemma 6.2.1 *If (C1)–(C3) are satisfied, then (6.1.1) possesses nonzero bounded on $\mathbb{R}$ solutions $\tilde{x}(t)$ which satisfy the following equations*

$$
\tilde{x}^{n-1}(t) = \begin{cases}
\dfrac{2\left(\displaystyle\int_{-\infty}^{\zeta_j} \Sigma(\zeta_j, s)2(n-1)q(s)ds\right)^{\frac{n}{2n-1}}}{\displaystyle\int_{-\infty}^{t} \Sigma(t,s)2(n-1)q(s)ds + \int_{-\infty}^{\zeta_j} \Sigma(\zeta_j, s)2(n-1)q(s)ds}, & \text{if } \alpha < 0, \\[2em]
-\dfrac{2\left(\displaystyle\int_{\zeta_j}^{\infty} \Sigma(\zeta_j, s)2(n-1)q(s)ds\right)^{\frac{n}{2n-1}}}{\displaystyle\int_{t}^{\infty} \Sigma(t,s)2(n-1)q(s)ds + \int_{\zeta_j}^{\infty} \Sigma(\zeta_j, s)2(n-1)q(s)ds}, & \text{if } \alpha > 0,
\end{cases}
$$

where $t \in [\theta_j, \theta_{j+1}]$.

Proof Consider $\alpha < 0$. First of all, let us show that

$$
\tilde{y}(t) = \int_{-\infty}^{t} \Sigma(t,s)2(n-1)q(s)ds
$$
$$
= \sum_{-\infty}^{k=j} Y(t, \theta_k) \int_{\zeta_{k-1}}^{\zeta_k} \Psi(\theta_k, s)2(n-1)q(s)ds + \int_{\zeta_j}^{t} \Psi(t, s)2(1-n)q(s)ds
$$

is a bounded solution of (6.2.2) for $t \in [\theta_j, \theta_{j+1}]$. Indeed,

$$
\tilde{y}'(t) = \sum_{-\infty}^{k=j} \big[2(1-n)p(t)Y(t, \theta_k)
$$
$$
+ 2(1-n)p(t)Y(\gamma(t), \theta_k)\big] \int_{\zeta_{k-1}}^{\zeta_k} \Psi(\theta_k, s)2(n-1)q(s)ds
$$
$$
+ \int_{\zeta_j}^{t} 2(1-n)p(t)\Psi(t, s)2(1-n)q(s)ds + \Psi(t, t)2(n-1)q(t)
$$
$$
= 2(1-n)p(t)\tilde{y}(t) + 2(1-n)p(t)\tilde{y}(\gamma(t)) + 2(n-1)q(t).
$$

Thus, $\tilde{y}(t)$ satisfies (6.2.2). It is easy to see that $\tilde{y}(t)$ is continuous in any interval (θ_i, θ_{i+1}), $i \in \mathbb{Z}$. We show that $\tilde{y}(t)$ is also continuous at points θ_i. Fix any $i \in \mathbb{Z}$.

$$
\tilde{y}(\theta_i+) = \sum_{-\infty}^{k=j} Y(\theta_i, \theta_k) \int_{\zeta_{k-1}}^{\zeta_k} \Psi(\theta_k, s)2(n-1)q(s)ds
$$
$$
+ \int_{\zeta_j}^{\theta_i} \Psi(\theta_i, s)2(1-n)q(s)ds
$$
$$
= \tilde{y}(\theta_i-)
$$

Next, we show that $\widetilde{y}(t)$ is bounded and separated from zero. One can confirm existence of positive numbers h and H such that $h \leq \|\Psi(t,s)\| \leq H$ for all $t, s \in [\theta_i, \theta_{i+1}]$, $i \in \mathbb{Z}$. Thus, we have that,

$$0 < \frac{2\underline{\theta} mhk(n-1)}{1 - e^{-\alpha\overline{\theta}}} + 2\underline{\theta} mh(n-1) \leq \|\widetilde{y}(t)\|$$
$$\leq \frac{4\overline{\theta} MKH(n-1)}{1 - e^{-\alpha\underline{\theta}}} + 4\overline{\theta} MH(n-1) < \infty,$$

for $t \in [\theta_j, \theta_{j+1}]$. On the other hand, by (C1) and (C3) one can verify that $\widetilde{y}(t) > 0$,, i.e., $\widetilde{y}(t)$ is separated from zero.

Finally, one can see that $y(t) + y(\gamma(t)) = 2x^{1-n}(t)x^{-n}(\gamma(t))$ implies $x(\gamma(t)) = y^{\frac{1}{2n-1}}(\gamma(t))$. Thus, the result follows from (C2) and the relation $x^{n-1}(t) = \dfrac{2y^{\frac{n}{2n-1}}(\gamma(t))}{y(t) + y(\gamma(t))}$.

We omit the proof of the case $\alpha > 0$, since it can be obtained in the similar manner. This finalizes the proof of lemma. □

Corollary 6.2.1 *If (C1)–(C3) are satisfied, then (6.1.1) possesses nonzero bounded on* $\mathbb{R}$ *solutions* $\overline{x}(t)$ *which satisfy the following equations*

$$\overline{x}^{n-1}(t) = \begin{cases} \dfrac{2\left(Y(\zeta_i, t_0)x_0^{-2m} + \int\limits_{t_0}^{\zeta_j} \Sigma(\zeta_j, s)2(n-1)q(s)ds \right)^{\frac{n}{2n-1}}}{\int\limits_{-\infty}^{t} \Sigma(t,s)2(n-1)q(s)ds + Y(\zeta_i, t_0)x_0^{-2m} + \int\limits_{t_0}^{\zeta_j} \Sigma(\zeta_j, s)2(n-1)q(s)ds}, & \text{if } \alpha < 0, \\[3em] \dfrac{2\left(Y(\zeta_i, t_0)x_0^{-2m} + \int\limits_{t_0}^{\zeta_j} \Sigma(\zeta_j, s)2(n-1)q(s)ds \right)^{\frac{n}{2n-1}}}{-\int\limits_{t}^{\infty} \Sigma(t,s)2(n-1)q(s)ds + Y(\zeta_i, t_0)x_0^{-2m} + \int\limits_{t_0}^{\zeta_j} \Sigma(\zeta_j, s)2(n-1)q(s)ds}, & \text{if } \alpha > 0, \end{cases}$$

where $t \in [\theta_j, \theta_{j+1}]$.

In what follows, we have different bifurcation scenarios depending on the parity of n. In the next sections, we deal with pitchfork and transcritical bifurcations, respectively.

6.3 The Pitchfork Bifurcation

Consider (6.1.1) for $n = 2m + 1$. That is,

$$x'(t) = p(t)x(t) - q(t)x^{2m+1}(t)x^{2m+1}(\gamma(t)). \tag{6.3.6}$$

Theorem 6.3.1 *Suppose that (C1)–(C3) are fulfilled for (6.3.6). Then, for $\alpha > 0$ the trivial solution is asymptotically stable whereas the nonzero bounded solutions $\overline{x}(t)$ are asymptotically unstable, and for $\alpha < 0$ the trivial solution is asymptotically unstable and the nonzero bounded solutions $\overline{x}(t)$ stable and $\widetilde{x}(t)$ are asymptotically pullback stable.*

Proof One can verify that the solution of (6.3.6) satisfies the following equation, [2, 4, 8, 16],

$$
x^{2m}(t, t_0, x_0) = \frac{2 \left(Y(\zeta_i, t_0) x_0^{-2m} + \int_{t_0}^{\zeta_i} \Sigma(\zeta_i, s) 4mq(s) ds \right)^{\frac{2m+1}{4m+1}}}{Y(t, t_0) x_0^{-2m} + \int_{t_0}^{t} \Sigma(t, s) 4mq(s) ds + Y(\zeta_i, t_0) x_0^{-2m} + \int_{t_0}^{\zeta_i} \Sigma(\zeta_i, s) 4mq(s) ds}.
$$
(6.3.7)

In the previous section, we have shown that (6.3.6) admits the trivial solution and bounded solutions $\widetilde{x}(t)$ and $\overline{x}(t)$. Asymptotic behavior of (6.1.1) depends on the sign of α. We start with the case $\alpha > 0$. One can confirm the following equation.

$$
x^{2m}(t, t_0, x_0) = \frac{2 \left(\widetilde{y}(\zeta_i) + Y(\zeta_i, t_0) \left(x_0^{-2m} - \widetilde{y}(t_0) \right) \right)^{\frac{2m+1}{4m+1}}}{\widetilde{y}(t) + \widetilde{y}(\zeta_i) + Y(t, t_0)(x_0^{-2m} - \widetilde{y}(t_0)) + Y(\zeta_i, t_0)(x_0^{-2m} - \widetilde{y}(t_0))}.
$$
(6.3.8)

From (6.3.8), it follows that $x^{2m}(t, t_0, x_0) \to 0$ as $t \to \infty$. So, $x(t, t_0, x_0) \to 0$ as $t \to \infty$, replying that all solutions are attracted forward to the point $\{0\}$. On the other hand, $x(t)$ converges to $\overline{x}(t)$ as $t \to -\infty$ whenever $||x_0|| < ||\widetilde{y}(t_0)||^{\frac{1}{2m}}$. Thus, the nonzero bounded solutions $\overline{x}(t)$ are asymptotically unstable.

If $\alpha < 0$, we notice that the expression (6.3.8) holds. Thus, one can see that $x(t)$ converges to $\widetilde{x}(t)$ as $t_0 \to -\infty$ and to $\overline{x}(t)$ as $t \to \infty$ whenever $||x_0|| < ||\widetilde{y}(t_0)||^{\frac{1}{2m}}$. Thus, $\widetilde{x}(t)$ is asymptotically pullback stable, whereas $\overline{x}(t)$ forward stable. Moreover, $x^{2m}(t, t_0, x_0) \to 0$ as $t \to -\infty$ whenever $||x_0|| < ||\widetilde{y}(t_0)||^{\frac{1}{2m}}$. Therefore, the origin is asymptotically unstable. The theorem is proved. $\square$

6.4 The Transcritical Bifurcation

In this section, we consider (6.1.1) for $n = 2m$. That is,

$$
x'(t) = p(t)x(t) - q(t)x^{2m}(t)x^{2m}(\gamma(t)).
$$
(6.4.9)

Theorem 6.4.1 *Suppose that (C1)–(C3) are fulfilled for (6.4.9). Then, for $\alpha > 0$ the trivial solution is asymptotically stable, and for $\alpha < 0$ the trivial solution is* *asymptotically unstable and the nonzero bounded solution $\overline{x}(t)$ is forward stable and $\widetilde{x}(t)$ pullback stable.*

Proof One can show that the solution of (6.4.9) satisfies the following equation, [2, 4, 8, 16],

$$x^{2m-1}(t, t_0, x_0) =$$

$$\frac{\left(2 \left(Y(\zeta_i, t_0) x_0^{-2m+1} + \int_{t_0}^{\zeta_i} \Sigma(\zeta_i, s) 2(2m-1) q(s) ds \right)^{\frac{2m}{4m-1}} \right)}{Y(t, t_0) x_0^{-2m+1} + \int_{t_0}^{t} \Sigma(t, s) 2(2m-1) q(s) ds + Y(\zeta_i, t_0) x_0^{-2m+1} + \int_{t_0}^{\zeta_i} \Sigma(\zeta_i, s) 2(2m-1) q(s) ds}. \qquad (6.4.10)$$

In Sect. 6.2, we have shown that (6.4.9) admits the trivial solution and the nonzero bounded solutions $\widetilde{x}(t)$ and $\overline{x}(t)$. As in Sect. 6.3, it is clear that asymptotic behavior of (6.4.9) depends on α. Consider the case $\alpha > 0$. From the Eq. (6.4.10), it follows that $x(t, t_0, x_0) \to 0$ as $t \to \infty$ as long as $x(\tau, t_0, x_0)$ exists for all $\tau \in [t_0, t]$. If $x_0 > 0$, observe that

$$Y(t, t_0) x_0^{-2m+1} + \int_{t_0}^{t} \Sigma(t, s) 2(2m - 1) q(s) ds + Y(\zeta_i, t_0) x_0^{-2m+1}$$

$$+ \int_{t_0}^{\zeta_i} \Sigma(\zeta_i, s) 2(2m - 1) q(s) ds > 0,$$

for $\tau \in [t_0, t]$. Thus, $x(\tau, t_0, x_0)$ exists for all $\tau \in [t_0, t]$ and does not blow up as $t \to \infty$.

If $x_0 < 0$, to ensure the existence of the solution $x(\tau, t_0, x_0)$, it is sufficient to show that

$$Y(\tau, t_0) x_0^{-2m+1} + \int_{t_0}^{\tau} \Sigma(\tau, s) 2(2m - 1) q(s) ds < 0,$$

for $\tau \in [t_0, t]$. Since $\int_{t_0}^{\tau} \Sigma(\tau, s) 2(2m - 1) q(s) ds > 0$, we require

$$|x_0| < \left(\frac{Y(\tau, t_0)}{\int_{t_0}^{\tau} \Sigma(\tau, s) 2(2m - 1) q(s) ds} \right)^{\frac{1}{2m-1}}.$$

However, one needs to show that right-hand side of the last inequality is bounded from below. One can find that

$$\frac{Y(\tau,t_0)}{\int_{t_0}^{\tau} \Sigma(\tau,s)2(2m-1)q(s)ds} =$$

$$\frac{1}{\int_{t_0}^{\xi_i} \Psi(\tau,s)2(n-1)q(s)ds + \sum_{k=i}^{k=j} Y(t_0,\theta_k)\int_{\xi_{k-1}}^{\xi_k} \Psi(\theta_k,s)2(n-1)q(s)ds + Y^-(\tau,t_0)\int_{\xi_j}^{\tau} \Psi(\tau,s)2(1-n)q(s)ds},$$

$$(6.4.11)$$

for $\theta_j \le \tau \le \theta_{j+1}$. It is easy to see that the last expression is bounded from below since $Y^{-1}(\tau, t_0)$ is bounded for a large enough τ.

Finally, we consider the case $\alpha < 0$. To show that the trivial solution is asymptotically unstable, notice that

$$x^{2m-1}(t, t_0, x_0) = \frac{2\left(\tilde{y}(\zeta_i) + Y(\zeta_i,t_0)\left(x_0^{-2m+1} - \tilde{y}(t_0)\right)\right)^{\frac{2m}{4m-1}}}{\tilde{y}(t) + \tilde{y}(\zeta_i) + Y(t,t_0)(x_0^{-2m+1} - \tilde{y}(t_0)) + Y(\zeta_i,t_0)(x_0^{-2m+1} - \tilde{y}(t_0))}. \quad (6.4.12)$$

From the last expression, it follows that $x(t)$ converges to 0 as $t \to -\infty$ for all $0 < x_0 < \tilde{y}^{\frac{1}{2m-1}}(t_0)$.

It remains to show that $\bar{x}(t)$ is forward and $\tilde{x}(t)$ pullback stable. If $x_0 > 0$, then it is clear that

$$Y(\tau, t_0)x_0^{-2m+1} + \int_{t_0}^{\tau} \Sigma(\tau, s)2(2m-1)q(s)ds > 0,$$

for $\tau \in [t_0, t]$. Thus, the solution $x(\tau, t_0, x_0)$ exists for all $\tau \in [t_0, t]$, and the (6.4.12) implies that $\bar{x}(t)$ is forward and $\tilde{x}(t)$ pullback stable for all $0 < x_0 < \tilde{y}^{\frac{1}{2m-1}}(t_0)$.

If $x_0 < 0$, then to ensure the existence of the solution $x(\tau, t_0, x_0)$, it is sufficient to show that

$$Y(\tau, t_0)x_0^{-2m+1} + \int_{t_0}^{\tau} \Sigma(\tau, s)2(2m-1)q(s)ds < 0,$$

for $\tau \in [t_0, t]$. Since $\int_{t_0}^{\tau} \Sigma(\tau, s)2(2m-1)q(s)ds > 0$, we require

$$|x_0| < \left(\frac{Y(\tau, t_0)}{\int_{t_0}^{\tau} \Sigma(\tau, s)2(2m-1)q(s)ds}\right)^{\frac{1}{2m-1}}.$$

The right-hand side of the last inequality is bounded from below because (6.4.11) holds. The theorem is proved.

6.5 Illustrative Examples

In this section, to illustrate theoretical results of Theorem 6.4.1, we consider two examples.

Example 1 Let us consider the following system.

$$x'(t) = (1.1 + \sin(5 + t^3/5))x(t) - (4 + 2.5\tanh(t/2))x^4(t)x^4(\gamma(t)), \quad (6.5.13)$$

where we have taken $p(t) = 1.1 + \sin(5 + t^3/5)$, $q(t) = 4 + 2.5\tanh(t/2)$, $\theta_k = \frac{k-1}{2}, k \in \mathbb{Z}$, $\zeta_k = \frac{k-1}{2}$, and $n = 4$. One can guarantee that $\alpha > 0$ and the conditions of Theorem 6.4.1 are satisfied with $m = 1.5$, $M = 6.5$, and $\underline{\theta} = \overline{\theta} = 1/2$. Theorem 6.4.1 guarantees that (6.5.13) has nonzero bounded solutions $\widetilde{x}(t)$ and $\overline{x}(t)$. Figure 6.1 reveals that all solutions starting near the origin diverge from the origin and converge to the nonzero bounded solutions. Therefore, the origin is asymptotically unstable and the bounded solutions are stable as expressed in the numerical simulations.

Example 2 We consider the following system.

$$x'(t) = -(1.1 + \sin(5 + t^3/5))x(t) - (4 + 2.5\tanh(t/2))x^4(t)x^4(\gamma(t)),$$
$$(6.5.14)$$

where for this example, we have taken $p(t) = -1.1 - \sin(5 + t^3/5)$, $q(t) = 4 + 2.5\tanh(t/2)$, $\theta_k = \frac{k-1}{2}, k \in \mathbb{Z}$, $\zeta_k = \frac{k-1}{2}$, and $n = 4$. One can guarantee that $\alpha <$

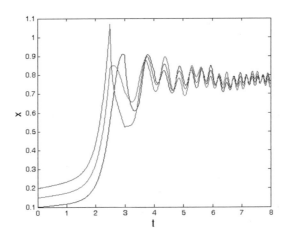

Fig. 6.1 Asymptotic behavior of (6.5.13) for $t \in [0, 8]$. In the figure, the *black color* corresponds to the solution with initial value $x_0 = 0.1$; the *red color* corresponds to the solution with initial value $x_0 = 0.15$; and the *blue color* corresponds to the solution with initial value $x_0 = 0.2$. One can see that all solutions which start in the neighborhood of the origin diverge from the origin and converge to the nontrivial bounded solutions $\overline{x}(t)$

Fig. 6.2 Asymptotic
behavior of (6.5.14) for
$t \in [0, 8]$. In the figure, the
black color corresponds to
the solution with initial value
$x_0 = 0.1$; the *red color*
corresponds to the solution
with initial value
$x_0 = -0.15$; and the *blue
color* corresponds to the
solution with initial value
$x_0 = 0.2$

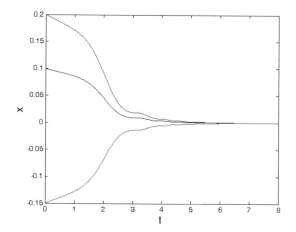

0 and the conditions of Theorem 6.4.1 are satisfied with $m = 1.5$, $M = 6.5$, and
$\underline{\theta} = \overline{\theta} = 1/2$. We present in Fig. 6.2 the solution of (6.5.14) with initial values $x_0 = -0.15, 0.1$ and $x_0 = 0.2$. Numerical simulations support our theoretical discussion and reveal that all solutions starting near the origin converge to the origin.

6.6 Notes

It is only in the recent decades there have been intensive developments on time-dependent differential equations. Local theory of dynamical systems is concerned with asymptotic behavior of an equilibrium or a periodic solution. However, in nonautonomous dynamical systems it is usually hard to find an equilibrium point or a periodic solution. Indeed, in many cases they fail even to exist. Therefore, equilibria generically persist as bounded solutions in the theory of time-varying dynamical systems. There are abstract formulation of a nonautonomous dynamical systems as new concept of nonautonomous attractors which are called pullback attractors [81, 83, 139, 155, 204, 227]. We investigate appearances and disappearances of bounded solutions that are stable and unstable in the pullback and forward sense. In particular, it was possible to study bifurcation analysis in nonautonomous systems with pullback attractors [75, 138, 152, 154]. In Chap. 4, we have defined an impulsive Bernoulli equation and studied nonautonomous transcritical and pitchfork bifurcations analysis depending on Lyapunov exponents [30, 32–34]. In the present chapter, we continue with scalar systems in hybrid dynamics and obtain results depending on the sign of these exponents. A theory of nonautonomous bifurcations in Banach space is treated in terms of exponential dichotomy in series of remarkable papers [198–200].

In the present chapter, it is the first time the Bernoulli equations with piecewise constant argument of generalized type have been studied. This chapter provides new sufficient conditions guaranteeing the existence and the separation from zero of the nonzero bounded solutions. Moreover, both forward and pullback asymptotic behavior of the trivial and the nonzero bounded solutions and different nonautonomous bifurcation scenarios are obtained.

Cooke, Shah, and Wiener were pioneers to initiate the theory of differential equations with piecewise constant argument [94, 221]. Since then, these equations have been under intensive investigations. The main idea of differential equation with piecewise constant argument is representing a hybrid of continuous and discrete dynamical systems and combining the properties of both the differential and difference equations. The concept of differential equations with piecewise constant argument has been generalized by introducing arbitrary piecewise constant functions as arguments in [2, 7, 8, 16]. The results obtained in [31] constitute the main part of this chapter.

References

1. M. Akhmet, *Principles of Discontinuous Dynamical Systems* (Springer, New-York, 2010)
2. M. Akhmet, *Nonlinear Hybrid Continuous/Discrete-Time Models* (Atlatntis Press, Paris, 2011)
3. M. Akhmet, E. Yilmaz, *Neural Networks with Discontinuous/Impact Activations* (Springer, New-York, 2013)
4. M.U. Akhmet, On the general problem of stability for impulsive differential equations. J. Math. Anal. Appl. **288**, 182–196 (2003)
5. M.U. Akhmet, On the smoothness of solutions of impulsive autonomous systems. Nonlinear Anal.: TMA **60**, 311–324 (2005)
6. M.U. Akhmet, Perturbations and Hopf bifurcation of the planar discontinuous dynamical system. Nonlinear Anal.: TMA **60**, 163–178 (2005)
7. M.U. Akhmet, Integral manifolds of differential equations with piecewise constant argument of generalized type. Nonlinear Anal.: TMA **66**, 367–383 (2007)
8. M.U. Akhmet, On the reduction principle for differential equations with piecewise constant argument of generalized type. J. Math. Anal. Appl. **336**, 646–663 (2007)
9. M.U. Akhmet, Almost periodic solutions of differential equations with piecewise constant argument of generalized type. Nonlinear Anal.: HS **2**, 456–467 (2008)
10. M.U. Akhmet, Asymptotic behavior of solutions of differential equations with piecewise constant arguments. Appl. Math. Lett. **21**, 951–956 (2008)
11. M.U. Akhmet, Stability of differential equations with piecewise constant argument of generalized type. Nonlinear Anal.: TMA **68**, 794–803 (2008)
12. M.U. Akhmet, Almost periodic solutions of the linear differential equation with piecewise constant argument. Discrete Impuls. Syst. Ser. A Math. Anal. **16**, 743–753 (2009)
13. M.U. Akhmet, Devaney's chaos of a relay system. Commun. Nonlinear Sci. Numer. Simul. **14**, 1486–1493 (2009)
14. M.U. Akhmet, Li-Yorke chaos in the impact system. J. Math. Anal. Appl. **351**, 804–810 (2009)
15. M.U. Akhmet, Shadowing and dynamical synthesis. Int. J. Bifurc. Chaos **19**(10), 1–8 (2009)
16. M.U. Akhmet, Exponentially dichotomous linear systems of differential equations with piecewise constant argument. Discontin. Nonlinearity Complex. **1**(4), 337–352 (2012)
17. M.U. Akhmet, D. Arugaslan, Bifurcation of a non-smooth planar limit cycle from a vertex. Nonlinear Anal.: TMA **71**, e2723–e2733 (2009)
18. M.U. Akhmet, D. Arugaslan, Lyapunov-Razumikhin method for differential equations with piecewise constant argument. Discrete Contin. Dyn. Syst. Ser. A **25**, 457–466 (2009)
19. M.U. Akhmet, D. Arugaslan, M. Beklioglu, Impulsive control of the population dynamics, in *Proceedings of the Conference on Differential and Difference Equations at the Florida*

© Springer Nature Singapore Pte Ltd. and Higher Education Press 2017
M. Akhmet and A. Kashkynbayev, *Bifurcation in Autonomous
and Nonautonomous Differential Equations with Discontinuities*,
Nonlinear Physical Science, DOI 10.1007/978-981-10-3180-9

Institute of Technology, August 1–5, 2005, Melbourne, Florida, ed. by R.P. Agarval, K. Perera (Hindawi Publishing Corporation, 2006), pp. 21–30

20. M.U. Akhmet, D. Arugaslan, M. Turan, Hopf bifurcation for a 3D Filippov system. Dyn. Contin. Discrete Impuls. Syst. Ser. A **16**, 759–775 (2009)

21. M.U. Akhmet, D. Arugaslan, X. Liu, Permanence of non-autonomous ratio-dependent predator-prey systems with piecewise constant argument of generalized type. Dyn. Contin. Discrete Impuls. Syst. Ser. A **15**, 37–52 (2008)

22. M.U. Akhmet, D. Arugaslan, E. Yilmaz, Stability analysis of recurrent neural networks with piecewise constant argument of generalized type. Neural Netw. **23**, 805–811 (2010)

23. M.U. Akhmet, D. Arugaslan, E. Yilmaz, Stability in cellular neural networks with piecewise constant argument. J. Comput. Appl. Math. **233**, 2365–2373 (2010)

24. M.U. Akhmet, D. Arugaslan, E. Yilmaz, Method of Lyapunov functions for differential equations with piecewise constant delay. J. Comput. Appl. Math. **235**, 4554–4560 (2011)

25. M.U. Akhmet, M. Beklioglu, T. Ergenc, V.I. Tkachenko, An impulsive ratio-dependent predator-prey system with diffusion. Nonlinear Anal. Real World Appl. **7**, 1255–1267 (2006)

26. M.U. Akhmet, G.A. Bekmukhambetova, A prototype compartmental model of the blood pressure distribution. Nonlinear Anal.: RWA **11**, 1249–1257 (2010)

27. M.U. Akhmet, C. Buyukadali, Periodic solutions of the system with piecewise constant argument in the critical case. Comput. Math. Appl. **56**, 2034–2042 (2008)

28. M.U. Akhmet, C. Buyukadali, Differential equations with a state-dependent piecewise constant argument. Nonlinear Anal.: TMA **72**, 4200–4211 (2010)

29. M.U. Akhmet, C. Buyukadali, T. Ergenc, Periodic solutions of the hybrid system with small parameter. Nonlinear Anal.: HS **2**, 532–543 (2008)

30. M.U. Akhmet, A. Kashkynbayev, *An Impulsive Bernoulli Equations: The Transcritical and the Pitchfork Bifurcations* (submitted)

31. M.U. Akhmet, A. Kashkynbayev, *Nonautonomous Bifurcation in Hybrid Systems* (submitted)

32. M.U. Akhmet, A. Kashkynbayev, Non-autonomous bifurcation in impulsive systems. Electron. J. Qual. Theory Differ. Equ. **74**, 1–23 (2013)

33. M.U. Akhmet, A. Kashkynbayev, Nonautonomous transcritical and pitchfork bifurcations in impulsive systems. Miskolc Math. Notes **14**, 737–748 (2013)

34. M.U. Akhmet, A. Kashkynbayev, Finite-time nonautonomous bifurcation in impulsive system, in *Proceedings of 10th QTDE, Electronic Journal of Qualitative Theory of Differential Equations* (2016)

35. M.U. Akhmet, M. Kirane, M.A. Tleubergenova, G.W. Weber, Control and optimal response problems for quasi-linear impulsive integro-differential equations. Eur. J. Oper. Res. **169**, 1128–1147 (2006)

36. M.U. Akhmet, H. Oktem, S.W. Pickl, G.-W. Weber, An anticipatory extension of Malthusian model, in CASYS'05-Seventh International Conference. AIP Conference Proceedings **839**, 26–264 (2006)

37. M.U. Akhmet, M. Tleubergenova, Asymptotic equivalence of the quasi-linear impulsive differential equation and the linear ordinary differential equation. Miskolc Math. Notes **8**, 117–121 (2007)

38. M.U. Akhmet, M. Turan, The differential equations on time scales through impulsive differential equations. Nonlinear Anal.: TMA **65**, 2043–2060 (2006)

39. M.U. Akhmet, M. Turan, Bifurcation of 3D discontinuous cycles. Nonlinear Anal.: TMA **71**, e2090–e2102 (2009)

40. M.U. Akhmet, M. Turan, Differential equations on variable time scales. Nonlinear Anal.: TMA **70**, 1175–1192 (2009)

41. M.U. Akhmet, M. Turan, Bifurcation in a 3D hybrid system. Commun. Appl. Anal. **14**, 311–324 (2010)

42. M.U. Akhmet, M. Turan, Bifurcation of discontinuous limit cycles of the Van der Pol equation. Math. Comput. Simul. **95**, 39–54 (2014)

43. M.U. Akhmet, E. Yilmaz, Impulsive Hopfield type neural network systems with piecewise constant argument. Nonlinear Anal.: Real World Appl. **11**, 2584–2593 (2010)

44. M.U. Akhmetov, Asymptotic representation of solutions of regularly perturbed systems of differential equations with a non classical right-hand side. Ukr. Math. J. **43**, 1298–1304 (1991)
45. M.U. Akhmetov, Periodic solutions of systems of differential equations with a non classical right-hand side containing a small parameter (Russian), in *TIC: Collection: Asymptotic Solutions of Non Linear Equations with Small Parameter* (Akad. Nauk Ukr. SSR, Inst. Mat., Kiev, UBL, 1991), pp. 11–15
46. M.U. Akhmetov, On the expansion of solutions to differential equations with discontinuous right-hand side in a series in initial data and parameters. Ukr. Math. J. **45**, 786–789 (1993)
47. M.U. Akhmetov, On the smoothness of solutions of differential equations with a discontinuous right-hand side. Ukr. Math. J. **45**, 1785–1792 (1993)
48. M.U. Akhmetov, On the method of successive approximations for systems of differential equations with impulse action at nonfixed moments of time (russian), Izv. Minist. Nauki Vyssh. Obraz. Resp. Kaz. Nats. Akad. Nauk Resp. Kaz. Ser. Fiz.-Mat. **1**, 11–18 (1999)
49. M.U. Akhmetov, R.F. Nagaev, Periodic solutions of a nonlinear impulse system in a neighborhood of a generating family of quasiperiodic solutions. Differ. Equ. **36**, 799–806 (2000)
50. M.U. Akhmetov, N.A. Perestyuk, On the almost periodic solutions of a class of systems with impulse effect (Russian). Ukr. Mat. Zh. **36**, 486–490 (1984)
51. M.U. Akhmetov, N.A. Perestyuk, Almost periodic solutions of sampled-data systems. Ukr. Mat. Zh. (Russian) **39**, 74–80 (1987)
52. M.U. Akhmetov, N.A. Perestyuk, On motion with impulse actions on a surfaces (Russian), Izv.-Acad. Nauk Kaz. SSR Ser. Fiz.-Mat. **1**, 111–114 (1988)
53. M.U. Akhmetov, N.A. Perestyuk, The comparison method for differential equations with impulse action. Differ. Equ. **26**, 1079–1086 (1990)
54. M.U. Akhmetov, N.A. Perestyuk, Asymptotic representation of solutions of regularly perturbed systems of differential equations with a non-classical right-hand side. Ukr. Math. J. **43**, 1209–1214 (1991)
55. M.U. Akhmetov, N.A. Perestyuk, Differential properties of solutions and integral surfaces of nonlinear impulse systems. Differ. Equ. **28**, 445–453 (1992)
56. M.U. Akhmetov, N.A. Perestyuk, Periodic and almost periodic solutions of strongly nonlinear impulse systems. J. Appl. Math. Mech. **56**, 829–837 (1992)
57. M.U. Akhmetov, N.A. Perestyuk, On a comparison method for pulse systems in the space $\mathbb{R}^n$. Ukr. Math. J. **45**, 826–836 (1993)
58. B.B. Aldridge, G. Haller, P.K. Sorger, D.A. Laffenburger, Direct Lyapunov exponent analysis enables parametric study of transient signalling governing cell behaviour. Syst. Biol. (IEE Proc.) **153**, 425–432 (2006)
59. A.A. Andronov, C.E. Chaikin, *Theory of Oscillations* (Princeton Uni. Press, Princeton, 1949)
60. A.A. Andronov, A.A. Vitt, C.E. Khaikin, *Theory of Oscillations* (Pergamon Press, Oxford, 1966)
61. L. Arnold, *Random Dynamical Systems* (Springer, Berlin, 1998)
62. J. Awrejcewicz, M. Feckan, P. Olejnik, On continuous approximation of discontinuous systems. Nonlinear Anal. **62**, 1317–1331 (2005)
63. J. Awrejcewicz, C.H. Lamarque, *Bifurcation and Chaos in Nonsmooth Mechanical Systems* (World Scientific, Singapore, 2003)
64. S. Banerjee, J. Ing, E. Pavlovskaia, M. Wiercigroch, R.K. Reddy, Invisible grazings and dangerous bifurcations in impacting systems: the problem of narrow-band chaos. Phys. Rev. E **79**, 037201 (2009)
65. N.N. Bautin, E.A. Leontovich, *Methods and Rules for the Qualitative Study of Dynamical Systems on the Plane (Russian)* (Nauka, Moscow, 1990)
66. M. di Bernardo, C.J. Budd, A.R. Champneys, P. Kowalczyk, *Piecewise-Smooth Dynamical Systems* (Springer, London, 2008)
67. M. di Bernardo, C.J. Budd, A.R. Champneys, P. Kowalczyk, A.B. Nordmark, G.O. Tost, P.T. Piiroinen, Bifurcations in nonsmooth dynamical systems. SIAM Rev. **50**(4), 629–701 (2008)
68. M. di Bernardo, A. Nordmark, G. Olivar, Discontinuity-induced bifurcations of equilibria in piecewise-smooth and impacting dynamical systems. Phys. D **237**, 119–136 (2008)

69. Ja. Bernoulli, Explicationes, annotationes et additiones ad ea quin actis superiorum annorum de curva elastica, isochrona paracentrica, velaria, hinc inde memorata, partim controversa leguntur; ubi de linea mediarum directionum, aliisque novis, Acta Eruditorum Dec (1695), pp. 537–553

70. R. Bellman, *Mathematical Methods in Medicine* (World Scientific, Singapore, 1983)

71. G.D. Birkhoff, *Dynamical Systems*, vol. 9 (American Mathematical Society Colloquium Publications, New York, 1927)

72. T.R. Blows, N.G. Lloyd, The number of limit cycles of certain polynomial differential equations. R. Soc. Edinb. Sect. **98A**, 215–239 (1984)

73. N.N. Bogolyubov, N.M. Krylov, *Introduction to Nonlinear Mechanics* (Acad. Nauk Ukrainy, Kiev, 1937)

74. S. Busenberg, K.L. Cooke, *Models of vertically transmitted diseases with sequential-continuous dynamics, in Nonlinear Phenomena in Mathematical Sciences* (Academic Press, New York, 1982), pp. 179–187

75. T. Caraballo, J.A. Langa, On the upper semicontinuity of cocycle attractors for non-autonomous and random dynamical systems. Dyn. Contin. Discrete Impuls. Syst. **10**(4), 491–513 (2003)

76. T. Caraballo, J.A. Langa, J.C. Robinson, A stochastic pitchfork bifurcation in a reaction-diffusion equation. Proc. R. Soc. A **457**, 2041–2061 (2001)

77. V. Carmona, E. Freire, E. Ponce, Limit cycle bifurcation in 3D continuous piecewise linear systems with two zones. Appl. Chua's Circuit Int. J. Bifurc Chaos **15**(10), 3153–3164 (2005)

78. S. Castillo, M. Pinto, Existence and stability of almost periodic solutions to differential equations with piecewise constant arguments, Electron. J. Differ. Equ. **58**, 15 pp. (2015)

79. A. Chavez, S. Castillo, M. Pinto, Discontinuous almost automorphic functions and almost automorphic solutions of differential equations with piecewise constant arguments, Electron. J. Differ. Equ. **56**, 13 pp. (2014)

80. A. Chavez, S. Castillo, M. Pinto, Discontinuous almost periodic type functions, almost automorphy of solutions of differential equations with discontinuous delay and applications, Electron. J. Qual. Theory Differ. Equ. **75**, 17 pp. (2014)

81. D.N. Cheban, *Global Attractors of Non-autonomous Dissipative Dynamical Systems* (World Scientific, Singapore, 2004)

82. D.N. Cheban, P.E. Kloeden, B. Schmalfuss, Pullback attractors in dissipative nonautonomous differential equations under discretization. J. Dyn. Differ. Equ. **13**, 185–213 (2001)

83. D.N. Cheban, P.E. Kloeden, B. Schmalfuss, The relationship between pullback, forwards and global attractors of nonautonomous dynamical system. J. Dyn. Differ. Equ. **13**(1), 185–213 (2001)

84. C.-Y. Cheng, Induction of Hopf bifurcation and oscillation death by delays in coupled networks. Phys. Lett. A **374**, 178–185 (2009)

85. V.V. Chepyzhov, M.I. Vishik, Attractors of non-autonomous dynamical systems and their dimension. J. Math. Pures Appl. **73**, 279–333 (1994)

86. W. Chin, E. Ott, H.E. Nusse, C. Grebogi, Grazing bifurcations in impact oscillators, Phys. Rev. E **3**(50) (1994)

87. K.-S. Chiu, M. Pinto, Variation of parameters formula and Gronwall inequality for differential equations with a general piecewise constant argument. Acta Math. Appl. Sin. Engl. Ser. **27**(4), 561–568 (2011)

88. K.-S. Chiu, M. Pinto, Oscillatory and periodic solutions in alternately advanced and delayed differential equations. Carpathian J. Math. **29**(2), 149–158 (2013)

89. S.-N. Chow, J.K. Hale, *Methods of Bifurcation Theory* (Springer, New York, 1982)

90. K.E.M. Church, R.J. Smith, Analysis of piecewise-continuous extensions of periodic linear impulsive differential equations with fixed, strictly inhomogeneous impulses. Dyn. Contin. Discrete Impuls. Syst. Ser. B: Appl. Algorithms **21**, 101–119 (2014)

91. E.A. Coddington, N. Levinson, *Theory of Ordinary Differential Equations* (McGraw-Hill, New York, 1955)

92. B. Coll, A. Gasull, R. Prohens, Degenerate Hopf bifurcations in discontinuous planar systems. J. Math. Anal. Appl. **253**, 671–690 (2001)
93. F. Colonius, R. Fabbri, R.A. Johnson, M. Spadini, Bifurcation phenomena in control flows. Topol. Methods Nonlinear Anal. **30**, 2007 (2007)
94. K.L. Cooke, J. Wiener, Retarded differential equations with piecewise constant delays. J. Math. Anal. Appl. **99**, 265–297 (1984)
95. W.A. Coppel, *Dichotomies in Stability Theory*, vol. 629 (Lecture Notes in Mathematics (Springer, Berlin, 1978)
96. H. Crauel, Random point attractors versus random set attractors. J. Lond. Math. Soc. **63**, 413–427 (2001)
97. H. Crauel, A. Debussche, F. Flandoli, Random attractors. J. Dyn. Differ. Equ. **9**, 397–441 (1997)
98. H. Crauel, F. Flandoli, Attractors for random dynamical systems. Prob. Theory. Relat. Fields **100**, 365–393 (1994)
99. C.M. Dafermos, An invariance principle for compact process. J. Differ. Equ. **9**, 239–252 (1971)
100. L. Dai, M.C. Singh, On oscillatory motion of spring-mass systems subjected to piecewise constant forces. J. Sound Vib. **173**, 217–232 (1994)
101. J.L. Daleckii, M.G. Krein, *Stability of Solutions of Differential Equations in Banach Spaces*, vol. 43, Translations of Mathematical Monographs (American Mathematical Society, Providence, 1974)
102. H. Dankowicz, A.B. Nordmark, On the origin and bifurcations of stick-slip oscillations. Phys. D **136**, 280–302 (2000)
103. J. Diblik, B. Moravkova, Solution of linear discrete systems with constant coefficients, a single delay and with impulses, in *Dynamical System Modeling and Stability Investigation* (2011), pp. 28–29
104. Q. Din, T. Donchev, A. Nosheen, M. Rafaqat, Runge-Kutta methods for differential equations with variable time of impulses. Numer. Funct. Anal. Optim. **36**(6), 777–791 (2015)
105. C. Ding, A predator-prey model with state dependent impulsive effects. Ann. Polon. Math. **111**(3), 297–308 (2014)
106. K.G. Dishlieva, A.B. Dishliev, Uniformly finally bounded solutions to systems of differential equations with variable structure and impulses, Electron. J. Differ. Equ. **205**, 9 pp. (2014)
107. A. Domoshnitsky, M. Drakhlin, E. Litsyn, Nonoscillation and positivity of solutions to first order state-dependent differential equations with impulses in variable moments. J. Differ. Equ. **228**, 39–48 (2006)
108. R. Fabbri, R.A. Johnson, F. Mantellini, A nonautonomous saddle-node bifurcation pattern. Stoch. Dyn. **4**(3), 335–350 (2004)
109. M. Feckan, Bifurcation of periodic and chaotic solutions in discontinuous systems. Arch. Math. (Brno) **34**, 73–82 (1998)
110. M. Feckan, Dynamical systems with discontinuities. Dyn. Contin. Discrete Impuls. Syst. Ser. A Math. Anal. **16**, 789–809 (2009)
111. M. Feckan, M. Pospisil, Bifurcation of sliding periodic orbits in periodically forced discontinuous systems. Nonlinear Anal.: Real World Appl. **1**, 150–162 (2013)
112. M.I. Feigin, Doubling of the oscillation period with C-bifurcations in piecewise-continuous systems. J. Appl. Math. Mech. **34**, 822–830 (1970)
113. M.I. Feigin, On the structure of C-bifurcation boundaries of piecewise-continuous systems. J. Appl. Math. Mech. **42**, 820–829 (1978)
114. R.J. Field, E. Koros, R.M. Noyes, Oscillations in chemical systems. II. Through analysis of temporal oscillations in the bromate-cerium-malonic acid system. J. Am. Chem. Soc **94**, 8649 (1972)
115. A.F. Filippov, *Differential Equations with Discontinuous Right-hand Sides* (Kluwer Academic Publishers, Dordrecht, 1988)
116. M.H. Fredriksson, A.B. Nordmark, Bifurcations caused by grazing incidence in many degree of freedom impact oscillators. Proc. R. Soc. A **453**, 1261–1275 (1997)

117. M. Frigon, D. O'Regan, Impulsive differential equations with variable times. Nonlinear Anal.: TMA **26**, 1913–1922 (1996)
118. H. Fujii, T. Yamada, Phase difference locking of coupled oscillating chemical systems. J. Chem. Phys **69**, 3830 (1978)
119. C. Glocker, F. Pfeiffer, Multiple impacts with friction in rigid multibody systems. Nonlinear Dyn. **4**, 471–497 (1995)
120. A. Grudzka, S. Ruszkowski, Structure of the solution set to differential inclusions with impulses at variable times, Electron. J. Differ. Equ. **114**, 16 pp. (2015)
121. J. Guckenheimer, P.J. Holmes, *Nonlinear Oscillations, Dynamical Systems and Bifurcations of Vector Fields* (Springer, New-York, 1083)
122. I. Gyori, F. Hartung, On numerical approximation using differential equations with piecewise-constant arguments. Period. Math. Hungar. **56**(1), 55–69 (2008)
123. A. Halanay, D. Wexler, *Teoria calitativă a sistemelor cu impulsuri* (Editura Academiei Republicii Socialiste România, Bucuresti, 1968)
124. J.K. Hale, *Asymptotic Behavior of Dissipative Systems*, vol. 25 (Mathematical Surveys and Monographs (American Mathematical Society, Providence, 1988)
125. J.K. Hale, H. Kocak, *Dynamics and Bifurcations* (Springer, New York, 1991)
126. P. Hartman, *Ordinary Differential Equations* (Wiley, New York, 1964)
127. C. Henry, Differential equations with discontinuous right-hand side for planning procedure. J. Econom. Theory **4**, 541–551 (1972)
128. P.J. Holmes, The dynamics of repeated impacts with a sinusoidal vibrating table. J. Sound Vib. **84**, 173–189 (1982)
129. E. Hopf, Abzweigung einer periodishen Losung von einer stationaren Losung eines Differential systems, Ber. Math.-Phys. Sachsische Academie der Wissenschaften,
130. F.C. Hoppensteadt, E.M. Izhikevich, *Weakly Connected Neural Networks* (Springer, New York, 1997)
131. S.C. Hu, V. Lakshmikantham, S. Leela, Impulsive differential systems and the pulse phenomena. J. Math. Anal. Appl. **137**, 605–612 (1989)
132. G. Iooss, D.D. Joseph, *Bifurcation of Maps and Applications* (Springer, New York, 1980)
133. A.P. Ivanov, Impact oscillations: Linear theory of stability and bifurcations. J. Sound Vib. **178**(3), 361–378 (1994)
134. Q. Jiangang, F. Xilin, Existence of limit cycles of impulsive differential equations with impulses at variable times. Nonlinear Anal.: TMA **44**, 345–353 (2001)
135. R.A. Johnson, F. Mantellini, A nonautonomous transcritical bifurcation problem with an application to quasi-periodic bubbles. Discrete Contin. Dyn. Syst. **9**(1), 209–224 (2003)
136. A. Kelley, The stable, center-stable, center, center-unstable, unstable manifolds. J. Differ. Equ. **3**, 546–570 (1967)
137. P.E. Kloden, Pullback attractors in nonautonomous difference equations. J. Differ. Equ. Appl. **6**(1), 91–102 (2000)
138. P.E. Kloeden, Pitchfork and transcritical bifurcation in systems with homogeneous nonlinearities and an almost periodic time coefficients. Commun. Pure Appl. Anal. **3**, 161–173 (2004)
139. P.E. Kloeden, M. Rasmussen, *Nonautonomous Dynamical Systems* (AMS, Providence, 2011)
140. P.E. Kloeden, B. Schmalfuss, Asymptotic behavior of non autonomous difference inclusion. Sys. Control Lett. **33**, 275–280 (1998)
141. P.E. Kloeden, S. Siegmund, Bifurcations and continuous transitions of attractors in autonomous and nonautonomous systems. Int. J. Bifurc. Chaos **5**(2), 1–21 (2005)
142. P.E. Kloeden, D. Stonier, Cocycle attractors of nonautonomously perturbed differential equations. Dyn. Discrete Contin. Impuls. Syst. **4**, 221–226 (1998)
143. A.N. Kolmogorov, On the Skorokhod convergence (Russian). Teor. Veroyatn. i Prim. **1**, 239–247 (1956)
144. N.M. Krylov, N.N. Bogolyubov, *Introduction to Nonlinear Mechanics* (Acad. Nauk Ukrainy, Kiev, 1937)

145. S.G. Kryzhevich, Grazing bifurcation and chaotic oscillations of vibro-impact systems with one degree of freedom. J. Appl. Math. Mech. **72**(4), 383–390 (2008)
146. M. Kunze, *Non-Smooth Dynamical Systems*, vol. 1744 (Lecture Notes in Mathematics (Springer, Berlin, 2000)
147. M. Kunze, T. Küpper, Qualitative bifurcation analysis of a non-smooth friction-oscillator model. Z. Angew. Math. Phys. **48**, 87–101 (1997)
148. Y.A. Kuznetsov, *Elements of Applied Bifurcation Theory, Applied Mathematical Sciences* (Springer, New York, 1995)
149. Y.A. Kuznetsov, S. Rinaldi, A. Gragnani, One-parametric bifurcations in planar Filippov systems. Int. J. Bifurc. Chaos **13**, 2157–2188 (2003)
150. T. Küpper, S. Moritz, Generalized Hopf bifurcation for non-smooth planar systems. Philos. Trans. R. Soc. Lond. **359**, 2483–2496 (2001)
151. J.A. Langa, J.C. Robinson, A finite number of point observations which determine a non-autonomous fluid flow. Nonlinearity **14**, 673–682 (2001)
152. J.A. Langa, J.C. Robinson, A. Suarez, Stability, instability and bifurcation phenomena in non-autonomous differential equations. Nonlinearity **15**, 1–17 (2002)
153. J.A. Langa, J.C. Robinson, A. Suarez, Forwards and pullback behavior of non-autonomous Lotka-Volterra system. Nonlinearity **16**, 1277–1293 (2003)
154. J.A. Langa, J.C. Robinson, A. Suarez, Bifurcation in non-autonomous scalar equations. J. Differ. Equ. **221**, 1–35 (2006)
155. J.A. Langa, A. Suarez, Pullback performance for non-autonomous partial differential equations. Electron. J. Differ. Equ. **2002**(72), 1–20 (2002)
156. V. Lakshmikantham, D.D. Bainov, P.S. Simeonov, *Theory of Impulsive Differential Equations* (World Scientific, Singapore, 1989)
157. V. Lakshmikantham, S. Leela, S. Kaul, Comparison principle for impulsive differential equations with variable times and stability theory. Nonlinear Anal.: TMA **22**, 499–503 (1994)
158. V. Lakshmikantham, X. Liu, On quasistability for impulsive differential equations. Nonlinear Anal.: TMA **13**, 819–828 (1989)
159. R.I. Leine, H. Nijmeijer, *Dynamics and Bifurcations of Non-Smooth Mechanical Systems*, Lecture Notes in Applied and Computational Mechanics (Springer, Berlin, 2004)
160. R.I. Leine, D.H. Van Campen, B.L. Van de Vrande, Bifurcations in nonlinear discontinuous systems. Nonlinear Dyn. **23**, 105–164 (2000)
161. R.I. Leine, N. Van de Wouw, Stability properties of equilibrium sets of non-linear mechanical systems with dry friction and impact. Nonlinear Dyn. **51**(4), 551–583 (2008)
162. Y. Li, C. Wang, Three positive periodic solutions to nonlinear neutral functional differential equations with parameters on variable time scales. J. Appl. Math. **2012**, 516476 (2012)
163. X. Li, J. Wu, Stability of nonlinear differential systems with state-dependent delayed impulses. Autom. J. IFAC **64**, 63–69 (2016)
164. L. Liu, J. Sun, Existence of periodic solution for a harvested system with impulses at variable times. Phys. Lett. A **360**, 105–108 (2006)
165. X. Liu, R. Pirapakaran, Global stability results for impulsive differential equations. Appl. Anal. **33**, 87–102 (1989)
166. J. Llibre, E. Ponce, Bifurcation of a periodic orbit from infinity in planar piecewise linear vector fields. Nonlinear Anal. **36**, 623–653 (1999)
167. A.C.J. Luo, *Discontinuous Dynamical Systems on Time-varying Domains, Monograph Series in Nonlinear Physical Science* (Springer- HEP, New York, 2010)
168. A.C.J. Luo, *Discontinuous Dynamical Systems, Monograph Series in Nonlinear Physical Science* (Springer-HEP, New York, 2011)
169. A.M. Lyapunov, *The General Problem of the Stability of Motion* (Mathematical Society of Kharkov, Kharkov, 1892). (in Russian)
170. A.M. Lyapunov, *Sur les figures dequilibre peu differentes des ellipsodies dune masse liquide homogène donnee dun mouvement de rotation* (Academy of Science St, Petersburg, St. Petersburg, 1906). (in French)

171. J.E. Marsden, M. McCracken, *The Hopf Bifurcation and Its Applications*, vol. 19 (Applied Mathematical Sciences (Springer, New York, 1976)
172. J.L. Massera, J.J. Schaffer, *Linear Differential Equations and Function Spaces* (Academic Press, New York, 1966)
173. N. Minorsky, *Nonlinear Oscillations* (D. Van Nostrand Company Inc, Princeton, 1962)
174. N.M. Murad, A. Celeste, Linear and nonlinear characterization of loading systems under piecewise discontinuous disturbances voltage: analytical and numerical approaches, in *Power Electronics Systems and Applications* (2004), pp. 291–297
175. Y. Muroya, Persistence, contractivity and global stability in logistic equations with piecewise constant delays. J. Math. Anal. Appl. **270**, 602–635 (2002)
176. A.D. Myshkis, A.M. Samoilenko, Systems with impulses at fixed moments of time (Russian). Math. Sb. **74**, 202–208 (1967)
177. R.F. Nagaev, Periodic solutions of piecewise continuous systems with a small parameter (russian). Prikl. Mat. Mech. **36**, 1059–1069 (1972)
178. R.F. Nagaev, D.G. Rubisov, Impulse motions in a one-dimensional system in a gravitational force field. Sov. Appl. Mech. **26**, 885–890 (1990)
179. A.H. Nayfeh, B. Balachandran, *Applied Nonlinear Dynamics* (Wiley, New York, 1995)
180. V.V. Nemytskii, V.V. Stepanov, *Qualitative Theory of Differential Equations* (Princeton University Press, Princeton, 1966)
181. A.B. Nordmark, Non-periodic motion caused by grazing incidence in an impact oscillator. J. Sound Vib. **145**, 279–297 (1991)
182. C. Nunez, R. Obaya, A non-autonomous bifurcation theory for deterministic scalar differential equations. Discrete Contin. Dyn. Syst. (Ser. B) **9**, 701–730 (2008)
183. H.E. Nusse, J.A. Yorke, Border-collision bifurcations including "period two to period three" for piecewise smooth systems. Phys. D **57**, 39–57 (1992)
184. H.E. Nusse, J.A. Yorke, Border-collision bifurcations for piecewise smooth one-dimensional maps. Int. J. Bifurc. Chaos Appl. Sci. Eng. **5**, 189–207 (1995)
185. T. Pavlidis, A new model for simple neural nets and its application in the design of a neural oscillator. Bull. Math. Biophys. **27**, 215–229 (1965)
186. T. Pavlidis, Stability of a class of discontinuous dynamical systems. Inf. Control **9**, 298–322 (1966)
187. T. Peacock, J. Dabiri, Introduction to focus issue: Lagrangian coherent structures. Chaos **20**(1), 017501 (2010)
188. O. Perron, Uber Stabilitat und asymptotisches Verhalten der Integrale von Differentialgleichungssystemen. Math. Z. **29**, 129–160 (1928)
189. O. Perron, Die Stabilitatsfrage bei Differentialgleichungen. Math. Z. **32**, 703–728 (1930)
190. F. Pfeiffer, Multibody systems with unilateral constraints (russian). J. Appl. Math. Mech. **65**, 665–670 (2001)
191. M. Pinto, Asymptotic equivalence of nonlinear and quasi linear differential equations with piecewise constant arguments. Math. Comput. Model. **49**, 1750–1758 (2009)
192. M. Pinto, Cauchy and Green matrices type and stability in alternately advanced and delayed differential systems. J. Differ. Equ. Appl. **17**(2), 235–254 (2011)
193. M. Pinto, G. Robledo, Controllability and observability for a linear time varying system with piecewise constant delay. Acta Appl. Math. **136**, 193–216 (2015)
194. V. Pliss, A reduction principle in the theory of stability of motion. Izv. Akad. Nauk SSSR Ser. Mat. **28**, 1297–1324 (1964)
195. H. Poincaré, *Sur les propriés des fonctions définies par les équations aux différences partielles* (Thése, Gauthier-Villars, Paris, 1879). (in French)
196. H. Poincaré, Sur lequilibre dune masse fluids animes dun mouvement de rotation. Acta Math. **7**, 259–380 (1885). (in French)
197. H. Poincaré, *Les méthodes nouvelles de la mécanique céleste, 2, 3* (Gauthier-Villars, Paris, 1892)
198. C. Pötzsche, Nonautonomous bifurcation of bounded solutions I: A Lyapunov-Schmidt approach. Discrete Contin. Dyn. Syst. Ser. B **14**(2), 739–776 (2010)

199. C. Pötzsche, Nonautonomous bifurcation of bounded solutions II: a shovel-bifurcation pattern. Discrete Contin. Dyn. Syst. Ser. A **31**(3), 941–973 (2011)
200. C. Pötzsche, Nonautonomous bifurcation of bounded solutions III: crossing-curve situations. Stoch. Dyn. **12**(2), 1150017 (2012)
201. I. Rachunkova, J. Tomecek, A new approach to BVPs with state-dependent impulses. Bound. Value Probl. **2013**(22), 133 pp. (2013)
202. I. Rachunkova, J. Tomecek, Fixed point problem associated with state-dependent impulsive boundary value problems. Bound. Value Probl. **2014**, 172 (2014)
203. I. Rachunkova, J. Tomecek, Existence principle for BVPS with state-dependent impulses. Topol. Methods Nonlinear Anal. **44**(2), 349–368 (2014)
204. M. Rasmussen, *Attractivity and Bifurcation for Nonautonomous Dynamical Systems* (Springer, Berlin, 2007)
205. M. Rasmussen, Finite-time attractivity and bifurcation for nonautonomous differential equations. Differ. Equ. Dyn. Syst. **18**, 57–78 (2010)
206. K. Rateitschak, O. Wolkenhauer, Thresholds in transient dynamics of signal transduction pathways. J. Theor. Biol. **264**, 334–346 (2010)
207. C. Robinson, *Dynamical Systems: Stability, Symbolic Dynamics, and Chaos* (CRC Press, Boca Raton, 1995)
208. V.F. Rozhko, Lyapunov stability in discontinuous dynamic systems (russian). Differ. Equ. **11**, 761–766 (1975)
209. T. Rui-Lan, C. Qing-Jie, L. Zhi-Xin, Hopf bifurcations for the recently proposed smooth-and-discontinuous oscillator. Chin. Phys. Lett. **27**(7), 074701 (2010)
210. R.J. Sacker, Existence of dichotomies and invariant splittings for linear differential systems IV. J. Differ. Equ. **27**, 106–137 (1978)
211. R.J. Sacker, G.R. Sell, Existence of dichotomies and invariant splittings for linear differential systems I. J. Differ. Equ. **15**, 429–458 (1974)
212. R.J. Sacker, G.R. Sell, Existence of dichotomies and invariant splittings for linear differential systems II. J. Differ. Equ. **22**, 478–496 (1976)
213. R.J. Sacker, G.R. Sell, Existence of dichotomies and invariant splittings for linear differential systems III. J. Differ. Equ. **22**, 497–522 (1976)
214. A.M. Samoilenko, N.A. Perestyuk, Stability of solutions of impulsive differential equations (Russian). Differentsial'nye uravneniya **13**, 1981–1992 (1977)
215. A.M. Samoilenko, N.A. Perestyuk, *Differential Equations with Impulsive Actions (Russian)* (Vishcha Shkola, Kiev, 1987)
216. A.M. Samoilenko, N.A. Perestyuk, *Impulsive Differential Equations* (World Scientific, Singapore, 1995)
217. G. Sansone, Sopra una equazione che si presenta nelle determinazioni della orbite in un sincrotrone. Rend. Accad. Naz. Lincei **8**, 1–74 (1957)
218. G. Sansone, R. Conti, *Nonlinear Differential Equations* (McMillan, New York, 1964)
219. G.R. Sell, Nonautonomous differential equations and dynamical systems. II. Limiting equations. Trans. Am. Math. Soc. **127**, 263–283 (1967)
220. G.R. Sell, *Topological Dynamics and Ordinary Differential Equations* (Van Nostrand Reinhold Mathematical Studies, London, 1971)
221. S.M. Shah, J. Wiener, Advanced differential equations with piecewise constant argument deviations. Int. J. Math. Sci. **6**, 671–703 (1983)
222. S.W. Shaw, P.J. Holmes, Periodically forced linear oscillator with impacts: chaos and long-period motions. Phys. Rev. Lett. **51**, 623–626 (1983)
223. S.W. Shaw, P.J. Holmes, A periodically forced piecewise linear oscillator. J. Sound Vib. **90**, 129–155 (1983)
224. D.J.W. Simpson, *Bifurcations in Piecewise-Smooth Continuous Systems* (World Scientific, Singapore, 2010)
225. D.J.W. Simpson, J. Meiss, Andronov-Hopf bifurcations in planar, piecewise-smooth, continuous flows. Phys. Lett. A **371**(3), 213–220 (2007)

226. A.V. Skorokhod, Limit theorems for random processes, (Russian), Teor. Veroyatnost. i Primenen. 289–319 (1956)
227. B. Schmalfuss, Attractors for non autonomous dynamical system, in *Proceedings of Equadiff*, vol. 99 (Berlin) ed. by B. Fiedler, K. Grger, J. Sprekels (World Scientific, Singapore, 2000), pp. 684–689
228. C. Sun, M. Han, X. Pang, Global Hopf bifurcation analysis on a BAM neural network with delays. Phys. Lett. A **360**, 689–695 (2007)
229. J.M.T. Thompson, H.B. Stewart, *Nonlinear Dynamics and Chaos* (Wiley, England, 2002)
230. A.S. Vatsala, J. Vasundara Devi, Generalized monotone technique for an impulsive differential equation with variable moments of impulse. Nonlinear Stud. **9**, 319–330 (2002)
231. T. Veloz, M. Pinto, Existence, computability and stability for solutions of the diffusion equation with general piecewise constant argument. J. Math. Anal. Appl. **426**(1), 330–339 (2015)
232. G.S. Whiston, Global dynamics of a vibro-impacting linear oscillator. J. Sound Vib. **118**, 395–429 (1987)
233. S. Wiggins, *Global Bifurcation and Chaos: Analytical Methods* (Springer, New York, 1988)
234. Y. Zhang, J. Sun, Stability of impulsive delay differential equations with impulses at variable times. Dyn. Syst. **20**, 323–331 (2005)
235. Z.T. Zhusubaliyev, E. Mosekilde, *Bifurcations and Chaos in Piecewise-Smooth Dynamical Systems* (World Scientific, Singapore, 2003)
236. Y. Zou, T. Küpper, Generalized Hopf bifurcation emanated from a corner for piecewise smooth planar systems. Nonlinear Anal. **62**, 1–17 (2005)
237. Y. Zou, T. Küpper, W.-J. Beyn, Generalized Hopf bifurcation for planar Filippov systems continuous at the origin. J. Nonlinear Sci. **16**, 159–177 (2006)

Printed in the United States
By Bookmasters